Tumor Associated Markers

Tumor Associated Markers

THE IMPORTANCE OF IDENTIFICATION IN CLINICAL MEDICINE

Edited by

Paul L. Wolf, M.D.

Professor of Pathology
University of California, San Diego
La Jolla, California

Masson Publishing USA, Inc.

New York • Paris • Barcelona • Milan • Mexico City • Rio de Janeiro

Copyright © 1979, by Masson Publishing USA, Inc.

All rights reserved. No part of this book may be reproduced in any form, by photostat, microform, retrieval system, or any other means, without the prior written permission of the publisher.

ISBN: 0-89352-065-9

Library of Congress Number: 79-87540

Printed in the United States of America

DEDICATION

To all those devoted physicians
who perpetually attempt to relieve
the suffering of the patient
afflicted with malignancy.

".... *Cancer the Crab lies so still that you might think he was asleep if you did not see the ceaseless play and winnowing motion of the feathery branches round his mouth. That movement never ceases. It is like the eating of a smothering fire into rotten timber in that it is noiseless and without haste.*"

Rudyard Kipling

PREFACE

This book is an outgrowth of a medical symposium sponsored by the University of California Extension, Berkeley, San Francisco, and San Diego, held on the Berkeley campus, December 2-3, 1978. The purpose of the symposium was to update current information relating to the importance of identifying and monitoring tumor-associated markers in clinical medicine. An in-depth analysis of the production of various tumor antigens, hormones and enzymes by malignant cells was presented by expert investigators concerned with this important endeavor.

Current information suggests that most human malignancies have tumor-associated antigens which are released into the circulation and which may be measured by sensitive techniques such as competitive binding radioimmunoassay. Two of the antigens that have been intensively investigated are the carcinoembryonic antigen (CEA) and the alphafetoprotin (AFP). In addition to the discussions of the clinical importance of detecting and monitoring CEA and AFP at the meeting, a number of investigations were presented analyzing the biochemical structure of CEA and the relationship of alphafetoprotein and the immune response.

Ectopic production of hormones and various polypeptides by neoplastic cells of non-endocrine origin was presented, and the diagnostic and therapeutic relevance of this phenomenon was discussed. In addition, a thorough analysis of tumor-associated enzymes was presented. These enzymes including the Regan isoenzyme of alkaline phosphatase, lactic dehydrogenase, histaminase, muramidase, acid phosphatase, beta glucuronidase, creatine kinase, the recently described galactosyltransferase II isoenzyme and the ultrafast alkaline phosphatase. New techniques for identifying and isolating cancer antigens, especially from malignancies such as breast cancer, were described.

The major goal of the symposium and this book is to provide the physician with current information regarding tumor-associated markers relating to the presence of a malignant lesion and to follow the remission or relapse of the malignancy during the therapy of the lesion. The symposium highlighted the practical clinical application of procedures for identifying circulating tumor products in the diagnosis and treatment of cancer.

A short paper session occurred after the main program in which a number of investigators attending the symposium presented their current work. These papers are also included in this book.

Paul L. Wolf, M.D.

ACKNOWLEDGMENTS

I gratefully acknowledge the excellent secretarial work of Ms. Mary Oberg who typed and assisted in the editing of the entire manuscript.

I also owe a great debt to Nathan Cohen, Ph.D., Director, Curriculum Development in Science, Division of Letters and Sciences University Extension, University of California, Berkeley, who coordinated the program and to Jeanne Rich and Alan Frankenfield Jr. for their support and advice during the preparation of this manuscript.

CONTRIBUTORS

Balinsky, Doris, Ph.D.
    Department of Biochemistry, Iowa State University, Ames, Iowa.
    Cayanis, Efitihia, Cummins, Roger, South African Institute of Medical Research, Johannesburg, South Africa.
    Hammond, Kathryn D., Department of Chemical Pathology, Saint Mary's Hospital Medical School, London, England.

Burchiel, Scott W., Ph.D.
    Research Assistant Professor, University of New Mexico, College of Pharmacy, Albuquerque, New Mexico.
    Rubin, Mark, B.S., Giorgi, Janis, Ph.D., University of New Mexico School of Medicine, Immunobiology Laboratories.
    Peake, Glenn T., M.D., Department of Medicine, University of New Mexico.
    Warner, Noel L., Ph.D., Professor, University of New Mexico School of Medicine, Immunobiology Laboratories.
    (University of New Mexico Medical Center, Albuquerque, New Mexico)

Fishman, William H., Ph.D.
    President, La Jolla Cancer Research Foundation, La Jolla, California.

Goldberg, David M., M.D., Ph.D.
    Professor and Chairman, Department of Clinical Biochemistry, University of Toronto, and Biochemist-in-Chief, the Hospital for Sick Children, Toronto, Canada.

Ghosh, Bimal C., M.D.
    Associate Chairman, Division of Surgical Oncology, Cook County Hospital, Chicago, Illinois.
    Ghosh, Luna C., M.D., Associate Professor, Department of Pathology, Cook County Hospital, Chicago, Illinois.

Koett, John, M.D., Ph.D.
    Associate Director of Chemistry, U.S. Naval Regional Medical Center, San Diego, California.
    Howell, James, M.D., Head, Clinical Pathology, Department of Laboratory Medicine, U.S. Naval Regional Medical Center, San Diego, California.
    Wolf, Paul L., M.D., Professor of Pathology, University of California, San Diego, School of Medicine, La Jolla, California.

Rich, Marvin, Ph.D.
    Executive Vice-president and Scientific Director, Michigan Cancer Foundation, Meyer L. Prentis Cancer Center, Detroit, Michigan.
    Brennan, Michael· J., M.D., FACP, President and Medical Director Michigan Cancer Foundation, Mayer L. Prentis Cancer Center, Detroit, Michigan.

Sell, Stewart, M.D.
Professor of Pathology, University of California, San Diego, School of Medicine, La Jolla, California.

Shively, John E., M.D.
Director of Immuno-Chemistry, Division of Immunology, City of Hope National Medical Center, Duarte, California.

Todd, Charles W., Ph.D.
Chairman, Division of Immunology, City of Hope National Medical Center, Duarte, California.

Shuster, Joseph, M.D., Ph.D.
Professor of Medicine, Clinical Research Associate of National Cancer Institute of Canada.
Gold, P., O.C., M.D., Ph.D., F.R.S.(C)., Professor of Medicine, Associate of the Medical Research Council of Canada.
(McGill Cancer Centre, McIntyre Medical Sciences Building, Montreal, Quebec, Canada)

Smuckler, Edward A., M.D., Ph.D.
Professor and Chairman, Department of Pathology, University of California, San Francisco, Medical Center, San Francisco, California.

Tomasi, Thomas B., M.D., PH.D.
William H. Donner Professor of Immunology, Chairman, Department of Immunology, and Professor of Medicine, Mayo Medical School, Rochester, Minnesota.

Yoo, T. J., M.D., Ph.D.
Associate Professor of Medicine, Division of Allergy and Immunology, Department of Internal Medicine, the University of Iowa, Iowa City, Iowa.
Kuan, Kenneth, Vestling, Carl S., Kuo, Chao Y., Members, Research Program, Division of Allergy and Immunology, Departments of Medicine and Biochemistry, University of Iowa and VA Medical Centers, Iowa City, Iowa.

Yoo, T. J., M.D., PH.D.
Associate Professor of Medicine, Division of Allergy and Immunology, Department of Internal Medicine, the University of Iowa, Iowa City, Iowa.
Kuo, C., Patil, S., Kim, U., Ackerman, L., Cancilla, P., Chiu, H. C., Members, Research Program, Departments of Medicine, Pediatrics and Pathology, University of Iowa Hospitals and VA Medical Center, Iowa City, Iowa.

Weiser, Milton M., M.D.
Chief, Division of Gastroenterology and Nutrition, Department of Medicine, State University of New York, at Buffalo, Buffalo, New York.
Podolsky, Daniel, M.D., Intern, Massachusetts General Hospital, Boston, Massachusetts.

Wolf, Paul L., M.D.
Professor of Pathology, University of California, San Diego, School of Medicine, La Jolla, California.

TABLE OF CONTENTS

Speakers for the Plenary Session

The Importance of Identifying Tumor Markers P. Wolf	1
The Biology of Cancer E. Smuckler	20
Alphafetoprotein: Developmental, Diagnostic and Carcinogenic Implications S. Sell	25
Alpha-Fetoprotein and the Immune Response T. Tomasi	39
The Use of CEA as a Tumor Marker J. Shuster and P. Gold	52
The Structure of Carcinoembryonic Antigen and a Genetically Related Material J. Shively and C. Todd	62
Oncodevelopmental Isoenzymes W. Fishman	73
Enzymes in Human Cancer: From the Specific to the More General D. Goldberg	81
Cancer-associated Galactosyltransferase and Glycopeptide Acceptor Activities M. Weiser and D. Podolsky	117

Speakers for the Short Paper Session

Immunocytological Demonstration of Growth Hormone in Mammary Carcinoma Cells   130
   B. Ghosh and L. Ghosh

Prostaglandin Production by Murine Tumors   133
   S. Burchiel, M. Rubin, J. Giorgi, G. Peake, H. Warner

Effect of Glucocorticoids on the Production of Alpha-fetoprotein and Lactate and Malate Dehydrogenases by Hepatoma   145
   T. Yoo, K. Kuan, C. Vestling, C. Kuo

Control of Oncofetal Antigen (Alpha-fetoprotein) Production As An Epiphenomenon of Tumorigenesis   152
   T. Yoo, C. Kuo, S. Patil, U. Kim, L. Ackerman, P. Cancilla, and H. Chiu

Enzymes and Ioszymes in Human Hepatoma   165
   D. Balinsky, E. Cayanis, K. Hammond, R. Cummins

The Association of an Ultrafast Alkaline Phosphatase Isoenzyme With Malignancy   174
   J. Koett, J. Howell, P. Wolf

The Breast Cancer Prognostic Study: Analyzing the Metastatic Potential of Human Breast Cancers.   188

THE IMPORTANCE OF IDENTIFYING TUMOR MARKERS

Paul L. Wolf, M.D.

Professor of Pathology, University of California, San Diego, La Jolla, California 92037

Much enthusiasm exists to identify enzymes, hormones or polyamines which are produced by neoplastic cells and which may be useful to suggest the presence of a tumor and monitor the therapy of the neoplasm. These substances may be quantitated in the serum, urine, or pleural, pericardial or peritoneal effusions or the cerebrospinal fluid. It is essential to detect malignancies before they metastasize and thus it is important to ascertain the presence of a tumor by identifying markers and instituting therapy before metastasis occurs.

Tumor markers are frequently not specific. In addition, they may be associated with non-neoplastic diseases. It has been established that the phenotypic expression of a tumor marker varies from neoplastic cell to cell, and a variation in production of the marker exists during the natural progression of the malignancy. It should be emphasized that changing levels of the tumor marker may not mirror a change in the tumor burden. With therapy of the malignancy, a decrease in the quantity of a marker may signify that there is a response to the therapy. However, another possibility is that tumor still exists and has lost its ability to produce the marker.[1]

Relapse of a tumor is usually associated with the appearance of the marker. However, some neoplasms lose their ability to produce the marker and with recurrence, the marker may thus not be present. Thus, clinicians should be cautious in employing markers to detect neoplasms and measure their therapeutic response and realize that the exact significance of the marker is, at times, difficult to assess. Since tumor markers are important in identifying neoplasms and monitoring their treatment, they should be employed in spite of certain shortcomings.

ENZYMES ASSOCIATED WITH CANCER

An important group of substances produced by malignant cells are enzymes. Many enzymes are associated with the presence of cancer. Unfortunately an increased level of various serum enzymes in patients with cancer is frequently nonspecific and cannot be utilized to assess the presence of tumor nor can it be utilized to monitor a therapeutic response. An increased serum level of an enzyme associated with a malignancy is caused primarily by the synthesis of the enzyme by the neoplastic cell. In addition, destruction of tissues by the malignancy results in release of enzymes from the necrotic tissue.

The enzymes associated with cancer are listed in Tables 1 and 2. In general, enzymes associated with cancer are usually identified in serum. However, it is useful to search for the enzyme associated with cancer in other body fluids such as effusions or in urine. In addition, histochemical procedures on blood cells yield useful information relevant to the possible neoplastic nature or exact identity of the cells. For example, a leukocyte alkaline phosphatase stain performed on a peripheral blood smear will aid in differentiating a myeloid leukemoid reaction with an elevated score from chronic myelogenous leukemia which is associated with a low score.[2] It should be emphasized that conditions other than chronic myelogenous leukemia may also be associated with a low score (Table 3).

LACTIC DEHYDROGENASE

One of the earliest enzymes found to be associated with cancer was lactic dehydrogenase (LD). Unfortunately, LD is elevated in patients with cancer only when the malignancy is far advanced. Approximately fifty per cent of patients with far advanced cancer will exhibit an elevated serum LD derived from the release of LD from the proliferating malignant cells. In addition, with metastasis there may be destruction of various tissues and LD from these tissues contributes to the increased serum LD. When isoenzymes are determined, the LD associated with cancer is usually LD 4, 5. Other patterns which also may be present are (1) presence of all 5 isoenzymes, (2) LD 1 in malignant melanoma, (3) LD 2, 3 in leukemia and lymphoma, (4) LD 3 in carcinoma of the lung, (5) LD 4, 5 in GU malignancies and cancer involving the liver and pancreas.[3]

Patients with hepatomegaly and exhibiting an elevated ALP and LD with a normal bilirubin probably have a neoplastic infiltrate in the liver.[4] ALP and LD reflux into the serum in this condition with good excretion of bilirubin by unaffected liver cells. The increased serum LD is probably derived from the neoplastic cells and the affected hepatic cells. With successful therapy of the cancer, serum LD may return to normal.

Another useful test is to determine LD in an effusion. An exudate secondary to cancer, collagen disease or infection will be characterized by an LD higher than the serum LD, protein content greater than 2.5 gm/dl and glucose in fluid lower than that of serum. The infiltrating neoplastic or inflammatory cells or erythrocytes contribute to the high LD of the effusion.[5]

A transudate has an LD like that of serum, glucose like that of serum, and protein content of less than 2.5 gm/dl. Frequently the LD fluid/LD serum ratio is a better indicator than the protein content of the fluid to determine if the fluid is an exudate or a transudate.

ALKALINE PHOSPHATASE

Alkaline phosphatase (ALP) may be increased in patients with cancer. However, ALP is too nonspecific to be used as a cancer marker. When the level of serum ALP is found to be elevated, and it is not related to physiologic bone growth, one must consider that a pathologic lesion may be present in any of a number of various organs (Table 4). An elevated ALP level may consist of a combination of various ALP isoenzymes. For example, carcinomas may be widely metastatic to multiple organs, including liver and bone, which results in ALP isoenzymes from liver and bone and in ectopic production of ALP by the carcinoma.[6] Before one analyzes the various causes for an elevated ALP level, it is important to identify the specific organ lesions that are associated with an elevated ALP level other than cholestatic liver disease or osteoblastic bone lesions.

Another unusual cause for an elevated serum ALP level is ectopic production of alkaline phosphatase by a neoplasm. Fishman described Regan isoenzyme in a patient in whom a bronchogenic carcinoma developed.[7] The patient's serum contained an elevated ALP level that was extremely heat stable and inhibited by 0.05 M phenylalanine. Thus, those characteristics are similar to those exhibited by the placental isoenzyme. It also resembled the placental isoenzyme by electrophoresis, and another term for this isoenzyme is carcinoplacental isoenzyme. In addition to lung cancer, other malignant neoplasms have produced the enzyme, such as breast cancer and carcinoma of the colon. The presence of the enzyme may be used to monitor cancer therapy. If the cancer is successfully treated, Regan isoenzyme will disappear from the serum. The highest incidence in malignancy in women is found in ovarian cancer or other gynecological neoplasms such as carcinoma of the cervix. Regan isoenzyme may also be present in conditions that are predisposed to neoplasia, e.g., familial polyposis or ulcerative colitis.[8]

A variant of Regan isoenzyme is Nagao isoenzyme, which was named for the patient who had a metastatic carcinoma to the pleural surfaces. Adenocarcinoma of the pancreas or adenocarcinoma of the bile duct may produce this isoenzyme. Nagao isoenzyme is extremely heat stable and inhibited by phenylalanine. In addition, Nagao isoenzyme differs from Regan in that Nagao isoenzyme is also inhibited by L-leucine. It should be emphasized that the two major organs that, when diseased, lead to an elevated serum ALP level are the liver, including the biliary tract, and the skeletal system; these should be of prime consideration.

Various methods are available to identify the isoenzymes of ALP. The easiest laboratory test to perform is the heat-stability test. The serum is heated at 56° C for exactly 10 minutes. The extremely heat-labile isoenzymes (90% labile) are derived from bone, reticuloendothelial system and vascular endothelium. The extremely heat-stable (90% stable) isoenzymes are produced by the placenta and malignant cells (Regan and Nagao isoenzymes). The intermediate group of isoenzymes (60 to 80% stable) are derived from the liver and intestine. The intestinal, placental, and Regan isoenzymes are inhibited by phenylalanine, while the biliary, bone, and vascular isoenzymes are not.

The amino acid inhibitors of ALP prevent dephosphorylation of the enzyme substrates in which the amino acids form enzyme inhibitor substrate complexes and, thus, inhibit enzyme action.[9] Urea inactivation activity has characteristics similar to the heat test. The bone isoenzyme is inactivated by urea in contrast to the placental isoenzyme being resistant to urea inactivation. Other amino acids that are used in inhibition studies besides phenylalanine are L-leucine and homoarginine. Furthermore, electrophoresis of the isoenzymes yields important information. The liver phosphatase moves more rapidly toward the anode while the bone isoenzymes has a slower anodal mobility and may overlap the liver band. The intestinal isoenzyme moves slower than bone. The support media that are useful are acrylamide and cellulose acetate. Triton X-100 enhances separation.

In addition to the above techniques, other enzymes that are specific for the hepatobiliary tract should be used especially when ALP is elevated, and it is necessary to ascertain the cause for the ALP elevation. These are 5' nucleotidase, leucine aminopeptidase, and γ-glutamyl transferase.

5' NUCLEOTIDASE

In the liver, lung, brain and kidney, 5' nucleotidase (NTP, EC 3.1.3.5.) is found. It is located in the plasma membrane of hepatic parenchyma and bile ductular cells. This alkaline phosphatase hydrolyzes nucleotides with a phosphate radical attached to the 5' position. The substrate is adenosine-5'-monophosphate. Elevation of this enzyme level is associated with bile duct proliferation. The clinical significance of NTP elevation is that it is specific for hepatobiliary disease and also can be used for the detection of hepatic metastasis with or without jaundice. It is also elevated in primary biliary cirrhosis, drug-induced cholestasis, and extrahepatic obstructive lesions.[10]

LEUCINE AMINOPEPTIDASE

Leucine aminopeptidase (EC 3.4.11.2.) is present in the small intestine, renal tubules, liver, pancreas, testis and uterus. Leucine aminopeptidase is elevated especially with carcinoma of the head of the pancreas with or without metastases to the liver and cannot be used to differentiate intrahepatic from extrahepatic cholestasis.[11] Other causes for elevation of leucine aminopeptidase are acute pancreatitis, acute hepatitis and alcoholic cirrhosis, and pregnancy.

γ-GLUTAMYL TRANSFERASE

γ-glutamyl transferase (EC 2.3.2.2.) is present in the kidney, pancreas, liver, spleen, intestine, lung, heart, and brain. Normal serum activity is mostly derived from the liver. This enzyme is a sensitive indicator of liver disease, but it is not specific.[12] However, it is not related to physiologic bone growth or pregnancy. γ-glutamyl transferase is not a good enzyme to follow hepatic necrosis but correlates best with hepatobiliary-pancreatic duct obstruction--especially in patients with cancer. This enzyme is a good screening test for alcoholism--75% of chronic alcoholics have a serum elevation. γ-glutamyl transferase is seriously limited in its application because it is elevated with usage of various drugs, renal disease, cardiac disease, prostatic cancer metastatic to bone, and the postoperative state.[13] Drugs, such as phenobarbital and phenytoin (Dilantin), stimulate an increase in γ-glutamyl transferase.[14]

COPPER OXIDASE

In addition to ALP and LD, other enzymes, at times, are increased in malignancy. A sensitive indicator of relapse of a leukemia or lymphoma is the presence of an increased serum ceruloplasmin.[15] Ceruloplasmin is a combination of an α-2 globulin and copper. It is an oxidase and may be measured by its oxidative action on paraphenylaline diamine. An increase in ceruloplasmin is also present in other conditions, especially inflammatory, degenerative, or neoplastic lesions. It also increases in women taking oral contraceptives and in pregnancy.[16] It is obvious that an increase in ceruloplasmin is a sensitive laboratory indicator of activity of a malignancy (especially Hodgkin's disease or leukemia), but it is too nonspecific. In addition to an increase in serum copper in cancer, an increase in the copper content of certain malignancies has been identified.

CREATINE KINASE

Recently the ectopic production of creatine kinase BB (CK BB) by cancer cells has been investigated. It has recently been found that CK BB is increased in the serum in patients with carcinoma of the prostate, stomach and lung.[17] Thus, if a patient presents with an increased serum CK and the isoenzyme is BB, the clinician should consider first neurologic disease, smooth muscle damage, acute pulmonary infarction, renal failure or possibly carcinoma of prostate or lung or gastrointestinal tract.

HISTAMINASE

Two important tumor markers in the serum associated with medullary carcinoma of the thyroid are calcitonin and histaminase. Recently another malignancy, small cell carcinoma of the lung, has also caused an increase in serum calcitonin and histaminase.[18]

MURAMIDASE

An important serum or urine tumor marker in patients with acute leukemia is muramidase. This enzyme is especially produced by monoblasts. The highest serum or urine levels are found in acute monocytic leukemia.[19] Moderately elevated levels are found in acute myelomonocytic leukemia with low levels in acute lymphocytic leukemia. An increase in serum muramidase does not necessarily suggest that the patient has acute monocytic or myelomonocytic leukemia. Any condition associated with an increase in monocytes such as sarcoidosis or tuberculosis may also increase serum or urine muramidase.[20]

GALACTOSYLTRANSFERASE II

A cancer-associated galactosyltransferase isoenzyme (GT II) has been identified in serum and effusions. A group of 232 patients with cancer of 14 different tissue types were investigated. 71 per cent had measureable serum GT II activity. Colorectal, pancreatic and gastric carcinoma showed 73 per cent, 83 per cent, and 75 per cent respectively. Benign diseases such as alcoholic hepatitis and celiac disease also were associated with increased serum GT II. GT II was detected in effusions caused by malignancy. GT II may be a tumor product rather than a result of the host response.[21]

ACID PHOSPHATASE

One of the first enzymes to be used as a tumor marker was acid phosphatase. Most physicians utilize the serum acid phosphatase as a test for metastatic carcinoma of the prostate. The colorometric acid phosphatase test unfortunately is usually only increased with far advanced disease. Furthermore, increases in total serum acid phosphatase may be caused by a wide variety of various conditions (Table 4). The use of the tartrate inhibition test has permitted some differentiation between the various isoenzymes of acid phosphatase.[22] However, tartrate partially also inhibits the liver, spleen and kidney isoenzymes in addition to the prostatic isoenzyme (Table 5).

Since the colorometric test for serum acid phosphatase usually only becomes increased with advanced metastatic disease, it was important to develop a test for cancer of the prostate before it becomes far advanced. Recently Foti developed a solid phase radioimmunoassay for serum prostatic acid phosphatase.[23] The radioimmunoassay test diagnosed prostatic cancer in 33% Stage I, 79% Stage II, 71% Stage III and 92% Stage IV disease. Thus, this new method has identified a large number of patients in the early stages of the disease. The radioimmunoassay test will, if used as a screening test, hopefully lead to early diagnosis and treatment of this condition. The sensitivity of the test will lead to a few false positives caused by prostatic infarction and by biopsy of the prostate.[24]

ONCO-FETAL PROTEINS
Alpha-Fetoprotein

The two major onco-fetal markers are alphafetoprotein (AFP) and carcinoembryonic antigen (CEA).

AFP is an α-1-globulin produced by the liver and by germ cell gonadal neoplasms. A good correlation exists between the level of elevation and the tumor burden. With complete removal of the neoplasm, AFP levels will return to normal, and with relapse, AFP will again become elevated. Thus, serial determination of AFP must be performed if it is to be utilized as an effective tumor marker

A majority of patients suffering from primary hepatocarcinoma have marked elevations of AFP (frequently 1000 ng/ml). Thus, a patient presenting with prominent levels of AFP most likely has a primary carcinoma of the liver.[25] Although AFP is a sensitive indicator of hepatoma, it is nonspecific and is elevated in conditions other than hepatoma.

A rare malignancy which may cause marked elevations of AFP is hepatoblastoma. Slight to moderate increase in AFP occurs with carcinoma metastatic to the liver from such primary sites as lung, colon, esophagus, stomach, pancreas, and prostate. In addition, viral hepatitis and cirrhosis may cause AFP to be elevated.[26]

Extrahepatic neoplasms are also associated with elevation of AFP in the serum. The most common neoplasm associated with an increase in AFP is endodermal sinus tumor of the ovary or testis.[27]

Since AFP is normally produced by the fetal liver and appears during fetal life in fetal blood and the maternal circulation, it is a useful test in fetal abnormalities.[28] Good quantitation of normal AFP by RIA procedures has been established especially during the period of 14 to 18 weeks' gestation. If an increase in maternal blood AFP occurs during this period, the obstetrician suspects anencephaly, spina bifida, congenital nephrosis, tyrosinosis or intra-uterine hepatoblastoma.[29] Thus, maternal serum AFP serves as a marker for serious fetal abnormalities and suggests to the obstetrician that possible termination of the pregnancy should be initiated.

CARCINOEMBRYONIC ANTIGEN

Carcinoembryonic antigen was described by Gold and Freedman in 1965 and was identified as an antigen detected with a heterologous antiserum from extracts of adenocarcinoma of the colon and from extracts of the gastrointestinal tract of human embryos and fetuses.[30] CEA was originally found in tissues derived from endoderm but subsequently has been associated with neoplasms arising from mesoderm and ectoderm.[31]

The main malignancy in which CEA is clinically being utilized is colorectal carcinoma. A definite correlation exists between the quantity of CEA in the serum and the prognosis. A localized lesion to the mucosa Duke's A will not have as high a level as a Duke's carcinoma C.[32] Carcinoembryonic antigen (CEA) is a glycoprotein which is a constituent of the glycocalyx of embryonic entodermal epithelium. An adult type antigen replaces CEA in the fetal colon during the third trimester of pregnancy. If the patient with a colorectal malignancy has an elevated level before surgery, the serum CEA should become normal with complete removal of the cancer. However, if the CEA remains elevated postoperatively, incomplete removal or relapse of the carcinoma has occurred.

An important aspect of monitoring CEA in the postsurgical period is that a rising titer of CEA usually precedes the reappearance of symptoms by several months. A second look operation confirmed the presence of cancer in a majority of patients if CEA levels continued to rise following surgery.[33]

Elevation of serum CEA may be found in over half of the patients with other cancers especially of the thyroid, stomach, breast, ovary and uterine cervix and endometrium, lung and pancreas.

It is important to serially quantitate the CEA level since most carcinomas of entodermal origin produce serum levels of greater than 20 ng/ml. In contrast non-entodermal malignancies such as lymphoma and leukemia and sarcoma are associated with levels between 5 to 20 ng/ml. The presence of a slight elevation 5-20 ng/ml may also be present in approximately 25% of smokers and benign conditions listed in Table 6.[34]

Goldenberg recently has utilized an antibody to CEA to clinically localize metastatic cancer. Goats were immunized with CEA and an anti-CEA antibody was harvested. The anti-CEA antibody was labeled with radioiodine. The radioiodinated anti-CEA was injected into the patient with the possible metastatic cancer, and the malignant lesions were localized utilizing gamma ray scanners to detect anti-CEA in the CEA-producing neoplasms.[35] A number of cancer screening units now employ simultaneous multiple tumor markers in their programs. If these multiple tumor markers are utilized in screening programs to identify the presence of a cancer, approximately one-fourth of patients with an elevated CEA will be found to have a malignancy. Other tumor markers found to be useful in the cancer screening programs besides CEA are alphafetoprotein and human chorionic gonadotrophin.

ECTOPIC PRODUCTION OF HORMONES
Human Chorionic Gonadotrophin

The most established of the hormone tumor markers is human chorionic gonadotrophin (HCG). It serves as a reliable index for the presence of a hydatidiform mole or choriocarcinoma.[36]

Immunologic cross-reactivity with human luteinizing hormone exists. New radioimmunoassay procedures for the B subunit of HCG are now available for this biologically inactive, but hormone specific, component of HCG. These new procedures have improved the monitoring of these hormones in patients with trophoblastic tumors. RIA has also provided a means to detect the alpha subunit of HCG. At times this subunit may be elevated with metastatic disease when the total HCG or beta subunit have been normal. A wide variety of non-trophoblastic tumors may also have the capacity to produce HCG, 10 to 50 per cent of other malignancies including carcinomas of lung, stomach, pancreas, endometrium and hepatoma may ectopically produce HCG.[37]

ADRENOCORTICOTROPHIC HORMONE
Ectopic production of other hormones is also important in the

diagnosis and mangement of patients with cancer. One of the more important ones is ACTH. This hormone may be produced by undifferentiated carcinomas of the lung and produce a clinical Cushing's syndrome causing management problems of the associated hypokalemic metabolic alkalosis. Other tumors associated with ectopic ACTH production are, rarely, non-beta cell islet cell pancreatic tumors, thymoma and carcinoid neoplasms of lung and gastrointestinal tract.[38]

INAPPROPRIATE ADH

An interesting electrolyte abnormality which may be an early clue to the presence of undifferentiated carcinoma of the lung is marked hyponatremia caused by inappropriate ADH associated with the neoplasm. Cerebral edema may result with confusion and stupor. Carcinoma of the pancreas may also be associated with inappropriate ADH syndrome.[39]

INSULIN

Profound hypoglycemia may be the presenting sign of a neoplasm. The production of insulin is usually associated with a beta cell islet cell lesion of the pancreas. Non-beta islet cell lesions may produce either gastrin and cause the Zollinger-Ellison syndrome or peptides resulting in marked diarrhea (pancreatic cholera). Ectopic insulin production with hypoglycemia has also been demonstrated in bronchial carcinoids and carcinomas and marked hypoglycemia may also occur in association with large bulky tumors especially mesotheliomas, liposarcomas, and fibrosarcomas. The mechanism for the hypoglycemia in these tumors may be consumption of glucose by the neoplasm or production of insulin by the malignancy.[40]

ERYTHROPOIETIN

Polycythemia may be a presenting sign of various neoplasms due to production of erythropoietin by the tumor cells. The neoplasms which may produce erythrocytosis are carcinoma of the kidney, hepatoma, pheochromocytoma and cerebellar hemangioblastoma.[41]

CATECHOLAMINES

An important laboratory indicator for the diagnosis of pheochromocytoma is an increased quantity of catecholamines, metanephrine or vanillylmandelic acid (VMA). These may also be increased in neuroblastomas in addition to homovanillic acid (HVA).

PARATHORMONE

When a patient develops hypercalcemia, the workup should focus on hyperparathyroidism due to an adenoma or hyperplasia of the parathyroid glands or ectopic production of parathormone (PTH) by neoplasms such as carcinoma of kidney or squamous carcinomas of lung, esophagus, or anus. Hypercalcemia may, of course, results from bone destruction and liberation of calcium resulting from osteoclast activating factor (OAF) produced by the tumor cells.[42]

MONOCLONAL GAMMOPATHY

An important tumor marker relating to certain hematologic neoplasms is a monoclonal immunoglobulin protein associated with multiple myeloma or Waldenstrom's macroglobulinemia. Rarely, non-hematologic epithelial neoplasms may be associated with a monoclonal gammopathy due to a prominent immunologic response to the tumor. The most common lesions causing this pattern are carcinoma of colon, breast, or stomach.[43]

ESTROGEN RECEPTORS

Finally an important laboratory examination in assessing the response of breast cancer to both hormonal manipulation or chemotherapy is the estrogen receptor test. This marker does not relate to diagnosing the lesion but is important in the management of the patient.

When receptors are determined in breast cancer tissue, the estrogen receptor test should be performed on the primary instead of on metastatic tissue sites since there is higher positivity in the primary site.[44]

Premenopausal women have a lower positivity since endogenous estrogens may occupy the receptor sites. It has been demonstrated that there is a 50-70% favorable therapeutic response to endocrine manipulation if the breast cancer is estrogen receptor positive.[45] The response rate to chemotherapy was significantly higher in receptor positive tumors (86%) than in receptor negative tumors (30%).[46] It thus is important to accurately assess estrogen receptors. False positive and false negative tests do occur and are listed in Tables 8 and 9.

Other receptor sites are now being evaluated. The best endocrine manipulation response occurs when the breast cancer possesses both estrogen and progesterone receptors.[47] Approximately one-third of tumors studied were found to be positive for androgen receptors. Glucocorticoid and steroid receptors are now being evaluated especially in cancers of the breast, prostate, endometrium, and in patients with leukemia and melanoma.[48,49,50,51,52]

TABLE 1.   ENZYMES ASSOCIATED WITH CANCER

    Acid phosphatase   Muramidase
    Alkaline phosphatase   Galactosyl transferase
    Leucine aminopeptidase   Creatine kinase
    Gamma glutamyl transferase Beta glucuronidase
    Lactic dehydrogenase   Histaminase
    5' nucleotidase   Ceruloplasmin (copper oxidase)

TABLE 2.   ISOENZYMES ASSOCIATED WITH CANCER

    Carcinoplacental alkaline   Hairy cell leukemia acid
    phosphatase isoenzymes   phosphatase
        Regan isoenzyme   Isoenzymes of lactic dehydrog-
        Nagao isoenzyme   enase
    Prostatic carcinoma iso-   Galactosyl transferase II
        enzyme   Creatine kinase BB

TABLE 3.   LEUKOCYTE ALKALINE PHOSPHATASE

    <u>Elevated</u>   <u>Normal</u>
    Leukocytosis   Chronic lymphocytic leukemia
    Leukemoid reaction   Secondary polycythemia
    Acute lymphocytic leukemia Multiple myeloma
    Polycythemia vera
    Myeloproliferative dis-   <u>Low</u>
        orders   Chronic myelocytic leukemia
    Hodgkin's disease, active   Acute myelocytic leukemia
    Pregnancy   Acute monocytic leukemia
    Newborn infants   Infectious mononucleosis
    Acute hemorrhage   Paroxysmal nocturnal hemoglob-
                                      inuria
                                Congenital hypophosphatasia
                                Sarcoidosis

TABLE 4. PATHOLOGIC LESIONS OF VARIOUS ORGANS ASSOCIATED WITH ELEVATED SERUM ALP LEVEL

Liver
    Cholestatic lesions
Bone
    Osteoblastic lesions
Heart
    Organization of infarct
    Cardiac failure
Lung
    Organization of infarct
Pancreas
    Acute pancreatitis
Kidney
    Organization of acute infarction

Gastrointestinal
    Giant peptic ulcer
    Erosive or ulcerative lesion of small intestine in malabsorption
    Acute infarction of small intestine
    Erosive ulcerative lesions of colon
Spleen
    Organization of acute infarction
Neoplastic ectopic production
    Regan isoenzyme
    Nagao isoenzyme

TABLE 5. CAUSE FOR ELEVATED SERUM ACID PHOSPHATASE

1. Physiological
    Bone growth
    Pregnancy
2. Metastatic carcinoma of the prostate
3. Bone lesions, osteolytic and osteoblastic
    Hyperparathyroidism
    Paget's disease
    Primary and secondary bone cancer
4. Hepatic disease
    Metastatic cancer
    Viral hepatitis
    Cirrhosis
5. Myocardial infarction
6. Renal disease
7. Hematologic disorders
    Hemolytic anemia
    Multiple myeloma
    Thrombocytopenia
    Thromboembolism
8. Reticuloendoethelial disease
    Gaucher's disease
    Niemann-Pick disease
    Eosinophilic granuloma
    Histiocytic lymphoma
    Hodgkin's disease

TABLE 6. INHIBITION VS. NON-INHIBITION OF ISOENZYMES
          OF ACID PHOSPHATASE

	L-Tartrate	Formaldehyde	Ethyl Alcohol
Prostate	Inhibited	Not inhibited	Inhibited
Liver	Inhibited		Not inhibited
Kidney	Inhibited		
Spleen	Inhibited		Not inhibited
Gaucher's disease	Not inhibited		
Red blood cells	Not inhibited	Inhibited	Inhibited

TABLE 7. BENIGN CONDITIONS ASSOCIATED WITH SLIGHT ELEVATION-CEA

Pregnancy   Regional enteritis
Emphysema   Duodenal ulcer
Cirrhosis   Benign colorectal polyps
Ulcerative colitis   Fibroadenomas of breast
Diverticulitis

TABLE 8.   FALSE POSITIVE ESTROGEN RECEPTOR

Estrogen receptor positive tumor that does not respond to therapy.

Tumor heterogenity
Some cells positive--
Most cells negative.

Androgens work through another receptor site.
Step distal to binding is deranged.

TABLE 9.   FALSE NEGATIVE ESTROGEN RECEPTOR

Endocrine response in an estrogen receptor negative.

The receptor site destroyed by incorrect sample storage; normal tissue assayed; previous endocrine therapy; nuclear positivity instead of cytoplasm.

REFERENCES

1. Wolf, H.: Tumor-cell markers: A biologic shell game? NEJM July 20, 1978, pp. 146-147.

2. Wolf, P., Williams, D.: PRACTICAL CLINICAL ENZYMOLOGY, John Wiley Co., New York, p. 268-269.

3. Goldman, R.D., Kaplan, N.O., Hall, T.C.: Lactic dehydrogenase in human neoplastic tissues. Cancer Res., 24:389-399, 1964.

4. Betro, M.G.: Significance of increased alkaline phosphatase and lactate dehydrogenase activities coincident with normal serum bilirubin. Clin. Chem. 18:1427-1429, 1972.

5. Wroblewski, F., Wroblewski, R.: The clinical significance of lactic dehydrogenase activity of serous effusions. Ann. Int. Med. 48:813-822, 1958.

6. Stolbach, L., Krant, M., Fishman, W.: Ectopic production of alkaline phosphatase isoenzyme in patients with cancer. NEJM 281:757-762, 1969.

7. Fishman, W.: Perspectives on alkaline phosphatase isoenzymes. Am. J. Med. 56:617-650, 1974.

8. Nathanson, L., Fishman, W.: New observations on the Regan isoenzyme of alkaline phosphatase in cancer patients. Cancer 27:1388-1397, 1971.

9. Byers, D.A., Fernley, H.N., Walker, P.G.: Studies on alkaline phosphatase: Inhibition of human placental phosphoryl phosphatase by L-phenylalanine. Eur. J. Biochem. 29:197-204, 1972.

10. Hobbs, J.R., Campbell, D.M., Scheuer, P.J.: The clinical value of serum 5'-nucleotidase assay, in Wield, O. (ed): CLINICAL ENZYMOLOGY. Switzerland, S., Karger, vol. 2, 1966, p. 106.

11. Batsakis, J.G., Kremers, B.J., Thiessen, M.G., et al: Biliary tract enzymology: A comparison of serum alkaline phosphatase, leucine aminopeptidase and 5'-nucleotidase. Am. J. Clin. Pathol. 50:485-490, 1968.

12. Boone, D.J., Routh, J.I., Schrantz, R.: γ-Glutamyl transpeptidase and 5'-nucleotidase. Am. J. Clin. Pathol. 61:321-327, 1974.

13. Rosalki, S.B., Rau, D., Lehmann, D., et al: Determination of serum gamma-glutamyl transpeptidase activity and its clinical applications. Ann Clin. Biochem. 7:143-151, 1970.

14. Rosalki, S.B., Tarlow, D., Rau, D.: Plasma gamma-glutamyl transpeptidase elevation in patients receiving enzyme-inducing drugs. Lancet 2:376-377, 1971.

15. Ray, G.R., Wolf, P.L., Kaplan, H.S.: Value of laboratory indicators in Hodgkin's disease: Preliminary results: Natl. Cancer Inst. Monogr. 36:315-323, 1973.

16. Wolf, P., Enlander, D., Dalziel, J., Swanson, J.: Green plasma in blood donors. NEJM 281:204, 1969.

17. Feld, R.D., Witte, D.L.: Presence of creatine kinase BB in some patients with prostatic carcinoma. Clin. Chem. 23:1930-1932, 1977.

18. Baylin, S.B., Weisburger, W.R., Eggleston, J.C., Mendelsohn, G., Beaven, M.A., Abeloff, M.D., Ettinger, D.S.: Variable content of histaminase, L-dopa decarboxylase and calcitonin in small-cell carcinoma of the lung. NEJM 299:105-110, 1978.

19. Osserman, E.F., Lawlor, D.P.: Serum and urinary lysozyme (muramidase) in monocytic and myelomonocytic leukemia. J. Exp. Med. 124:921-952, 1966.

20. Perillie, P.E., Kaplan, S.S., Finch, S.C.: Significance of changes in serum muramidase activity in megaloblastic anemia. NEJM 277:10-12, 1967.

21. Podolsky, D.K., Weiser, M.M., Isselbacher, K.J., Cohen, A.M.: A cancer-associated galactosyltransferase isoenzyme. NEJM: 299:703-705, 1977.

22. Fishman, W.H., Bonnder, C.D., Hamburger, F.: Serum "prostatic" acid phosphatase and cancer of the prostate. NEJM 255:925-932, 1956.

23. Foti, A.G., Cooper, J.F., Herschman, H., Malvaez, R.: Detection of prostatic cancer by solid-phase radioimmunoassay of serum prostatic acid phosphatase. NEJM 297:1357-1361, 1977.

24. Gittes, R.: Acid phosphatase reappraised. NEJM 297:1398-1399, 1977.

25. Alpert, E., Hershberg, R., Schur, P.H., Isselbacher, K.J.: α-Fetoprotein in human hepatoma: Improved detection in serum and quantitative studies using a new sensitive technique. Gastroenterology 61:137-143, 1971.

26. Ruoslahti, E., Seppala, M., Rasanen, J.A., Vuopio, P., Helshe, J.: Alpha-fetoprotein and hepatitis B antigens in acute hepatitis and primary cancer of the liver. Scand. J. Gastroenterology. 8:1-6, 1973.

27. Kohn, J., Orr, A.H., McElwain, T.J., Bentali, M., Peckham, M.J.: Serum-alphafetoprotein in patients with testicular tumors. Lancet 2:433-436, 1976.

28. Laxova, R., Lewis, B.V., Suddaby, M.: A clinical service for prenatal diagnosis. Lancet, November 15, 1975, p. 964-966.

29. Leighton, P.C., Gordon, Y.B., Kitau, M.J., Leek, A.E., Chard, T.: Levels of alpha-fetoprotein in maternal blood as a screening test for fetal neural tube defect. Lancet, November 22, 1975, pp. 1012-1015.

30. Gold, P., and Freedman, S.O.: Specific carcinoembryonic antigens of the human digestive system. J. Exp. Med. 122:467-481, 1965.

31. Goldenberg, D.M., Sharkey, R.M., Primus, F.J.: Carcinoembryonic antigen in histopathology: Immunoperoxidase staining of conventional tissue sections. J. Nat. Cancer Inst. 57:11-22, 1976.

32. Zamchek, N.: The present status of CEA in diagnosis, prognosis and evaluation of therapy. Cancer 36:2460-2468, 1975.

33. Martin, E.W., James, K.K., Hurtubise, P.E., Cetalano, P., Minton, J.P.: The use of CEA as an early indicator for gastrointestinal tumor recurrence and second-look procedures. Cancer 39:440-446, 1977.

34. Hansen, H.J., Snyder, J.J., Miller, E., et al: Carcinoembryonic antigen (CEA) assay: a laboratory adjunct in the diagnosis and management of cancer. Hum. Pathol. 5:139-147, 1974.

35. Goldenberg, D.M., DeLand, F., Kim, E., Bennett, S., Primus, F.J., van Nagell, J.R., Estes, N., DeSimone, P., Rayburn, P.: Use of radiolabeled antibodies to carcinoembryonic antigen

for the detection and localization of diverse cancers by external photoscanning. NEJM, June 22, 1978, pp. 1384-1388.

36. Braunstein, G.D., Vaitukaitis, J.L., Carbone, P.P., Ross, G.T.: Ectopic production of human chorionic gonadotrophin by neoplasms. Ann. Intern. Med. 78:39-45, 1973.

37. Hattori, M., Fukase, M., Yoshimi, H., Matsukura, S., Imura, H.: Ectopic production of human chorionic gonadotropin in malignant tumors. Cancer 42:2328-2333, 1978.

38. Azzopardi, J.G., Williams, G.D.: Pathology of "non-endocrine" tumors associated with Cushing's syndrome. Cancer 22:274-286, 1968.

39. Schwartz, W.B., Bennett, W., Curelop, S., Bartter, F.C.: A syndrome of renal sodium loss and hyponatremia probably resulting from inappropriate secretion of anti-diuretic hormone. Amer. J. Med. 23:529-542, 1957.

40. Shames, J.M., Dhurandhar, N.R., Blackard, W.G.: Insulin-secreting bronchial carcinoid tumor with widespread metastasis. Amer. J. Med. 44:632-637, 1968.

41. Rosse, W.F., Waldemann, T.A.: A comparison of some physical and chemical properties of erythropoiesis-stimulating factors from different sources. Blood 24:739-749, 1964.

42. Omenn, G.S., Roth, S.I., Baker, W.H.: Hyperparathyroidism associated with malignant tumors of nonparathyroid origin. Cancer 24:1004-1012, 1969.

43. Hobbs, J.R.: Paraproteins, benign or malignant? Br. Med. J. 3:699-704, 1967.

44. Lippman, M.E., Allegra, J.C.: Estrogen receptor and endocrine therapy of breast cancer. NEJM 299:930-933, 1978.

45. Kiang, D.T., Kennedy, B.J.: Estrogen receptor assay in the differential diagnosis of adenocarcinomas. JAMA 238:32-34, 1977.

46. Kiang, D.T., Frenning, D.H., Goldman, A.I., Ascensao, V.F., Kennedy, B.J.: Estrogen receptors and responses to chemotherapy and hormonal therapy in advanced breast cancer. NEJM 299:1330-1334, 1978.

47. Persijn, J.P., Korsten, C.B., Engelsman, E.: Oestrogen and

androgen receptors in breast cancer and response to endocrine therapy. Br. Med. J. 4:503, 1975.

48. Maass, H., Engel, B., Trams, G., et al: Steroid hormone receptors in human breast cancer and the clinical significance. J. Steroid Biochem. 6:743-749, 1975.

49. Teulings, F.A.G., van Gilse, H.A.: Demonstration of glucocorticoid receptors in human mammary carcinomas. Hormone Res. 8:107-116, 1977.

50. Allegra, J.C., Lippman, M.E., Thompson, E.B., et al: Steroid hormone receptors in human breast cancer. Proc. Am. Soc. Cancer Res Am Soc. Clin. Oncol.19:336, 1978.

51. Lippman, M.E., Halterman, R.H., Leventhal, B.G., et al: Glucocorticoid binding proteins in acute lymphoblastic leukemic blast cells. J. Clin. Invest. 52:1715-1725, 1973.

52. Fisher, R.I. Neifeld, J.P. Lippman, M.E.: Oestrogen receptors in human malignant melanoma. Lancet 2:337-338, 1976.

THE BIOLOGY OF CANCER

Edward A. Smuckler, M.D., Ph.D.

Professor and Chairman, Department of Pathology, University of California Medical Center, San Francisco, California.

INTRODUCTION

The impact of cancer on man, in particular, and the very unusual biological nature of this disease have combined to stimulate a remarkable investment of time and effort in the scientific world, to understand the nature of the disease and to seek its cause. Unfortunately, the vast majority of the analyses that we have at hand are phenomenological. We have yet to be successful in addressing specific mechanisms in terms of molecular biology that permit us to describe the true nature of the disease. It is my intention to briefly discuss the biology of cancer, and to raise this to the phenotypic expression observed in several categories of neoplastic disease, a major thrust of the remainder of this symposium. It shall be my thesis that cancer represents an aberrant form of differentiation, a concept not new to biology, but an important one from the standpoint of our understanding of the pathology of the disease. It should be pointed out at the onset that the term "transformation" has sufficient imprecision to lead to some of the difficulties that occur in comparing discussions from different laboratories. I would like to suggest that mature end-stage cells are not transformed, but rather the process of neoplastic modification takes place during the period of development in which less mature, less differentiated cells alter their biological activity, a hypothesis discussed recently by Pierce, among others.[1,2]

What is Cancer? Neoplastic disease as defined by Willis is an uncoordinated growth of tissue unregulated by host factors that persists after the stimulus for such growth has ceased.[3] Simply stated, cancer seems to be a population explosion of one or more cell types. The characteristic feature that I would like to re-emphasize are that these represent a proliferation of cells, that their growth is uncoordinated with normal tissue, and most importantly that this growth persists after the stimulus that elicited it has ceased. From a classification standpoint, there are two types of neoplasms--benign and malignant (or cancerous). In the former instance, the proliferative response remains localized. The mass may grow to untoward proportions but unless the site of occurrence of the tumor is fortuitous, it will not in general kill the host. The characteristic features are continued proliferation of cells, uncoordinated by the host, continuing after the stimulus ceases. Importantly, the growth remains at the site of origin and does not invade adjacent tissues nor form colonies in distant sites.

In contrast, cancer cells have all of the same properties with one additional feature--the propensity to invade adjacent tissues and to spread to distant sites where nests of cells colonize and continue to grow. This is the biological *sine qua non* of the malignant or cancerous state. To summarize, both benign and malignant neoplasms represent uncoordinated proliferation of cells whose growth persists following cessation of the stimulus which elicits them while malignant cells have the unique propensity to invade adjacent structures and colonize distant sites. There are two salient points, a statement of fact and the other a question. The first is that all the properties of malignant cells are shared at some time during ontogeny. Embryonal cells proliferate following inciting stimuli, and embryonal cells are also able to invade and colonize distant sites. A critical difference concerns the regulation of these growths. The neoplasms with which we are concerned manifest unregulated growth. An important question concerns the relationship between malignant cells and benign tumor cells. From a morphological standpoint, there is a great deal of similarity in these growths. From a biological standpoint, these growths are uniquely different. Teleologically, it would be comfortable to make benign tumors only one step in the process toward malignant transformation. However, there is no evidence that the two processes are in any way related.

Origins of Spontaneous and Induced Malignant Neoplasms: Tissues have been classified on the basis of the residual mitotic capacity of the cell types involved.[4] Certain tissues and organs forfeit the capacity for cell division, a characteristic of such tissues as central nervous system neurons, and certain muscle cells. Other tissues continue to be maintained by a replicative pool, replacing lost cells during the normal process of turnovers. Examples of this are liver, endocrine tissue, and epithelium. If one tabulated the incidence of neoplastic disease in the human population, the vast majority are derived from those tissues which maintain a replicative pool with a few notable exceptions (Table I). It would seem that for the spontaneous neoplasms cell replication or the capacity for cell replication is an important prerequisite feature.

Properties of Neoplasms. It is particularly important that one attempt to separate properties of neoplastic growth which are unique to the cells involved and those that represent the host response to this aberrant proliferation. Clearly, the property of invasion and proliferation, and the presence of constitutive enzymes and cell markers are unique to the cells. On the other hand, vascularization of tumors, the fibrous response, sustenances of metastatic sites, and immune mechanisms brought to bear against tumors are host phenomena (Table II).

a. <u>Property of Malignant Tissue.</u> Probably one of the fundamental issues of malignant tissue is that the trait possessed by abnormal cells is a hereditable one. Malignant cells produce malignant cells. Furthermore, cancer cells seem to possess traits which are quantitatively different but not qualitatively different from the cells of origin of the neoplasm. This particular point is important, since it directs attention not to mature end-stage cells with which to compare the tumor but rather to the group of cells from which they arise. It thus appears that malignant neoplasms reflect no more than an altered phenotype, and one reminiscent of the developmental state. A case in point that reflects this type of change is the non-neoplastic transformation of the respiratory epithelium in response to injury. Cigarette smoking is associated with an altered phenotypic expression in epithelial cells in respiratory bronchus. Instead of the progeny of the reserve cells producing tall ciliated columnar epithelial cells, stratified keratinizing squamous cells are produced. This metaplastic phenomena however is a reversible one, and the altered differentiation of these cells may be reversed. By analogy, it would seem that if malignant transformation takes place from dividing cells, much of the expression we see may represent no more than an altered phenotype rather than the appearance of new properties in this cancer. Specifically, it is entirely reasonable to assume that the message for a variety of constitutive enzymes and surface phenomena found in less differentiated cells are more a property of the developmental state of its aberrant differentiation than any unique new feature.[5]

b. <u>Regulation of Phenotypic Expression.</u> Unfortunately, our understanding of regulation of phenotype in differentiation and development is limited. There are several responses which have been suggested to be regulatory phenomenon. First, the genome itself may have specific histone and non-histone proteins covering the genetic code which may regulate its expression. Secondly, there may be specific regulator codons which permit initiation of messenger formation in other sites. Third, there may be a selective capacity for the different RNA polymerases to read specific segments of the genome at any given time. Fourth, there is a processing of RNA within the nucleus which may have some regulatory phenomena selecting for or against specific messengers. Fifth, the transport of RNA from the nucleus to the cytoplasm is a regulated phenomena, and an energy dependent one capable of selection, and finally, it is entirely possible that there are cell cytoplasmic structures of functions which modify the ability to encode these messengers.[6] At present, we do not know which of these reactions is altered in the malignant cell.

There has been abundant investigation concerning the potential mutational events, either induced or secondary, as part of the transforming process. If indeed there is a mutational event involved, it must be specific and reproducible in different tissues and in different organisms to result in the similarity of cancers. Furthermore, evidence for alteration of base pair has been more tempered by the activity of repair enzymes, some of which may be specific for alkylated bases.[7]

Different histone-like proteins have not been regarded as a simple explanation for the altered phenotype. A greater similarity rather than a greater divergency of histone and non-histone proteins has been found in cells during development, differentiation and neoplasia. Attempts to identify regulatory systems for different reading by DNA dependent RNA polymerase is still in its infancy. The several problems attendant upon critical analysis of gene product under these circumstances has not been satisfactory.

There are two areas in which some suggestive evidence points to altered DNA formation or an altered code in the transforming process. In viral transformed cells, it is frequently possible to identify segments of viral genomes inserted into host nuclear material. Under these circumstances there is little difficulty in ascribing the source of the altered phenotype. The problems, however, are made evident by a consideration of the Epstein-Barr virus which appears in both neoplastic and non-neoplastic lymphoid cell lines depending upon the circumstances surrounding the host. The Burkitt's lymphoma and infectious mononucleosis must be contrasted to clarify this point.

The identification of altered gene product in neoplastic cells has been sought. The problems of hybridization are recognized but in point of fact no new products have been identified.[6] A gene dosage effect may be the critical feature. A more critical analysis of RNA processing and regulation has revealed in several induced tumor lines the capacity of the nucleus to sequester nuclear restricted RNA has been altered, and material previously suppressed appears in the cytoplasm.[8] This would provide a heuristic hypothesis for the appearance of constitutive enzymes and altered surface phenomena as well as the biological behavior in cancer cells. It simply directs our attention again to the fact that neoplastic cells represent an aberrant form of differentiation, rather than a different and unique process. It may also suggest that an understanding of regulatory phenomena in differentiation, and specifically epigenetic regulation, is a necessary and inherent part of the understanding of the malignant disease process.

REFERENCES

1. Pierce, G.B.: Fed. Proc. 29:1243, 1970.
2. Farber, E.: Cancer Research 33:2539, 1973.
3. Willis, R.A.: "Pathology of Tumors", 3rd ed., Butterworth and Co., Ltd. London, 1960.
4. Goss, R.J.: Science 153:1619, 1966.
5. Shearer, R.W., Smuckler, E.A.: Cancer Research 31:2104, 1970, Cancer Research 32:339, 1972.
6. Smuckler, E.A., Koplitz, M.: Cancer Research 34:827, 1974.
7. See discussion by I.B. Weinstein, Ann. N.Y. Acad. Sci. 271: 461, 1976.
8. Clawson, G., Smuckler, E.A.: Proc. Nat. Acad. Sci. 75:5400, 1978.

TABLE I. INCIDENCE OF SPONTANEOUS CANCERS IN ADULTS CONTRASTING NON-MITOTIC VS. MITOTIC TISSUES AND ORGANS

Non-Mitotic	Cancer Incidence	Mitotic	Cancer Incidence
Central Nervous System Neurons	Rare	Central Nervous System Supporting Cells	Not uncommon
Muscle Cells	Rare	Skin	Common
Adipose Cells	Rare	Breast Duct and Acinus	Common
		Pancreatic Duct	Common
		Pancreatic Acinus	Rare
		Liver	Not uncommon
		Small Bowel Epithelium	Rare
		Colon Epithelium	Common

TABLE II. BIOLOGICAL CANCER PROPERTIES

Intrinsic	Extrinsic
(Properties of cancer cells)	(Host responses to cancer cells)
Growth (proliferation)	Immunological response
Cell surface changes including "neoantigens"	Vascularization
Enzyme composition, including fetal and constitutive enzymes	Distant colonization (metastasis)
Invasion	

ALPHAFETOPROTEIN: DEVELOPMENTAL, DIAGNOSTIC AND CARCINOGENIC IMPLICATIONS

Stewart Sell, M.D.

Professor, Department of Pathology, University of California, San Diego, School of Medicine, La Jolla, California

INTRODUCTION

Alphafetoprotein (AFP) was discovered in 1963 by the Soviet scientist, Garri Israelevich Abelev. By immunodiffusion, he identified a protein present in fetal sera and in the sera of adult mice bearing a transplantable hepatocellular carcinoma that was not present in the sera of normal adults. Using more sensitive radioimmunoassays, elevated serum concentrations of AFP have been found not only in fetal, maternal, and hepatoma associated sera, but also following liver cell proliferation and after exposure to chemical hepatocarcinogens in the adult (Figure 1).

BIOCHEMICAL PROPERTIES OF AFP

The biochemical properties of rat AFP and albumin are listed in Table I. AFP and albumin are similar in molecular weight, have a reciprocal relationship in serum concentration during development and share up to 45% amino acid sequence homology. AFP is the major serum protein during fetal development but falls to barely detectable concentrations in the adult. On the other hand serum concentrations of albumin are low during fetal development, and increase as the concentration of AFP decreases to become the major sera protein in the adult. AFP of murine species specifically binds estrogen, whereas AFP of other species does not.

AFP DURING NORMAL DEVELOPMENT

AFP is made mainly by fetal liver and yolk sac (and to a small extent by fetal GI tract). AFP produced by the fetus is transferred rapidly across the placenta in most mammalian species, so that elevated serum concentrations of AFP occur in material sera during pregnancy. In the human, serum AFP concentrations fall during gestation. In the rat the serum AFP concentration does not fall until after birth and the fall coincides with a decline in the rate of hepatocellular proliferation.

AFP IN ABNORMAL DEVELOPMENT

Abnormalities in production or distribution of AFP during development may lead to high maternal blood or amniotic fluid concentrations of AFP (Table II). Measurement of maternal blood and amniotic fluid AFP concentrations have been found to be an effective method for the prenatal diagnosis of spinal cord defects and is used as a screening procedure in Great Britain for recommendations of therapeutic abortions.

POSSIBLE FUNCTIONS OF AFP

A listing of possible functions of AFP is shown in Table III. The most obvious function of AFP is that it simply serves as a fetal albumin. However, many investigators feel that AFP must also have a more specific function different than albumin. For instance, the fact that rat and mouse AFP bind estrogen has led to the idea that AFP protects the fetus from the effects of maternal estrogen. However, the AFP of other species does not bind estrogen, and even though AFP concentrations are high in the fetus, there is not nearly enough AFP available to neutralize the large amounts of maternal estrogen produced. A popular theory is that AFP protects the fetus from maternal immune rejection (Table IV). Whereas there is considerable evidence that AFP does have some effect on cultured lymphocytes *in vitro*, there are many observations that indicate that AFP is not specifically immunosuppressive (Table V). Finally, AFP may serve as a developmental tissue organizational signal. AFP is produced during normal development of the liver and during restitutive proliferation of the liver in the adult. At these times alignment of different proliferating cell types to produce hepatic lobular structures is required. Since other "oncodevelopmental" products such as murine T locus markers, lymphocyte differentiation markers (LY), placental hormones and isoenzymes have critical organizational functions, it is possible AFP may be required for normal alignment of different cell types in the liver.

PRODUCTION OF AFP BY TUMORS

Production of AFP by tumors in the adult is related to the tissue of origin; AFP-producing tumors usually arise from cells that produce AFP during fetal development (Table VI). Thus the production of a fetal-associated product by tumors is not a random property of any tumor. The production of fetal products by tumors reflects a level of developmental control that is still present in the tumor. Different levels of control are also found with other tumor-associated products such as CEA and ectopic hormones (Table VII).

PROGNOSTIC AND DIAGNOSTIC USE OF AFP

AFP production, as reflected in serum concentrations, may be used diagnostically and prognostically for patients with hepatocellular carcinomas. Factors affecting AFP production by transplantable rat hepatoms are listed in Table VIII. For each of these associations, there are notable exceptions. For instance, Morris hepatoma 9098 is fast growing but does not produce an elevation of serum AFP, hepatoma 7777 has only one extra chromosome, but produces high serum AFP, and poorly differentiated hepatomas produce a wide range of sera AFP concentrations. Recent studies in our laboratory using complementary DNA probes for AFP and albumin messenger RNA show that the ability of a given tumor to produce AFP or albumin is directly related to the amount of functional mRNA present. Thus control of AFP or albumin potential by a given tumor is pretranslational.

If a tumor produces AFP, the kinetics of the serum concentrations may be used to differentiate tumor production of AFP from that following restitutive liver proliferation and to evaluate the effects of tumor therapy. A steady increase in the serum concentration of AFP occurs during the growth of an AFP-producing tumor. Serum AFP elevations that occur following liver cell injury or after partial hepatectomy are transient and usually return to normal within one to two weeks. In an experimental model system, following of the serum concentration of AFP may be used to determine the effectiveness of surgery, irradiation or chemotherapy of transplantable hepatocellular carcinomas. Similar studies in humans with hepatocellular carcinoma or teratocarcinomas with yolk sac elements indicate that serum concentration of AFP may be used diagnostically and prognostically for these tumors.

AFP DURING CHEMICAL CARCINOGENESIS

The kinetics of serum AFP elevations, cellular alterations in the liver and the localization of AFP-containing cells during the exposure of rats to chemical carcinogens suggest that hepatocellular carcinomas may evolve by different pathways. Previously a sequential alteration of liver cells from a few cells with atypical staining characteristics, through foci of hyperplasia, to reversible neoplastic nodules was believed to be a common pathway induced by different carcinogens in development of malignant hepatocellular carcinomas. However a series of studies employing AFP as a marker for early events caused by hepatocarcinogens has revealed four major carcinogenic patterns (Table IX). Diethylnitrosamine (DEN) given orally at low, but carcinogenic doses produces no early AFP elevations or morphologic changes. However after a dose-dependent time period there is a rapid progressive increase in serum AFP associated with the appearance of multiple hepatocellular carcinomas. Diaminoazobenzene (DAB) produces an early necrosis of liver cells followed by hepatocellular proliferation, associated with a very slight temporary AFP elevation. This is followed by a prolonged rapid elevation of AFP, the appearance of neoplastic nodules, oval cells and finally hepatocellular carcinomas. Ethionine and N-2 fluorenylacetamide (FAA) produce no early morphologic change (or very little) but a prolonged moderate elevation of AFP. With the development of neoplastic nodules, the serum concentrations of AFP actually fall to almost normal. After prolonged exposure the serum concentrations of AFP may again become elevated if an AFP-producing tumor develops, but remain normal if a non-AFP-producing tumor appears. Azaserine and Wy-14643 (induces peroxisome proliferation) produce hepatocyte proliferation in the absence of necrosis within one to two days after administration. This is associated with a slight temporary elevation of AFP. Prolonged administration results in a decline of AFP to normal concentrations and the appearance of non-AFP-producing tumors.

Immunofluorescent examination of the liver during treatment with chemical carcinogens reveals three cell types containing AFP: newly-appearing bile duct-like cells, oval cells and rarely in zones of a glandular-like hyperplasia (atypical hyperplasia). Neoplastic nodules do not contain AFP positive cells. Thus a number of morphologic changes occur during exposure of rats to chemical carcinogens. Because many of the tumors produced synthesize AFP it is difficult to accept the conclusion that the nodules are the only precursor lesion. It is concluded that hepatocellular carcinomas induced by chemical carcinogens may arise from different cell types including neoplastic nodules, oval cells, zones of atypical hyperplasia, bile duct-like cells or even normal appearing hepatocytes.

SUMMARY

Alphafetoprotein is a serum protein with important implications for normal development, tumor diagnosis and chemical carcinogenesis. AFP is produced during gestation by fetal liver and yolk sac. High concentrations of AFP in amniotic fluid or maternal blood may indicate a serious congenital anomaly such as a neural tube defect. AFP has structural similarities to serum albumin and functions as a fetal albumin, but may also play a role in protecting the fetus from maternal immune rejection or as an organizational signal for development of hepatic lobules. AFP is produced by tumors in the adult that are derived from tissues that produce AFP during development, namely hepatocellular carcinomas and teratocarcinomas containing yolk sac elements, suggesting a retention of some level of developmental control of gene expression by tumors. Serum AFP concentrations may be used diagnostically to differentiate hepatomas from hepatitis and prognostically to follow the effect of therapy on AFP-producing tumors. In the rat model system, different carcinogens produce different patterns of AFP production prior to development of hepatocellular carcinomas suggesting multiple pathways of carcinogenesis. Thus studies on AFP have provided increased understanding normal development, liver cell injury, and induction and growth of tumors; but many critical questions remain unanswered.

ACKNOWLEDGEMENTS

The essential collaboration of F. F. Becker, B. Lombardi, H. Shinozuke, J. Reddy, H. L. Leffert, and H. Sela-repat is gratefully acknowledged. The technical assistance of D. Gord, H. Skelly, D. Stillman, M. Michaelson, K. Thomas, J. Scott, and C. Lloyd has contributed to various aspects of this work.

SELECTED REFERENCES

Abelev, G. I.: Alphafetoprotein in oncogenesis and its association with malignant tumors. Adv. Cancer Res. 14:295-385, 1971.

Sell, S.: The Biologic and Diagnostic Significance of Oncodevelopmental Gene Products IN H. Waters (Ed.) The Handbook of Cancer Immunology, New York: Garland Pub., In. Vol. 3, p. 1, 1978.

Sell, S. and Becker F. F.: Alphafetoprotein: Guest Editorial. J. Nat. Canc. Inst. 60:19-26, 1978.

Figure 1: Serum Concentrations of Alphafetoprotein in Rats

The serum concentration of AFP is over five logs higher than normal (0.03 µg/ml in newborn and tumor-bearing rats. Elevations also occur following liver injury or hepatocarcinogen exposure.

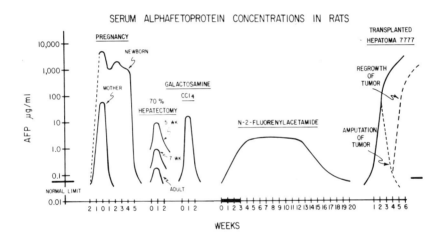

TABLE I. SOME PROPERTIES OF RAT AFP AND ALBUMIN

	AFP	ALBUMIN
MOLECULAR WEIGHT	70,000	60,000
ISOELECTRIC POINT	4.9	5.9
CHO CONTENT	4%	NONE
SERUM CONCENTRATION (ng/ml)		
FETAL	8,000	<100
ADULT	0.03	60,000
ESTROGEN BINDING		
K_a	10^{-9}	10^{-5}

TABLE II. INCREASED AFP CONCENTRATIONS IN ABNORMAL PREGNANCIES

ABNORMALITIES	MECHANISM
FETAL DEATH	RELEASE OF FETAL AFP
NEURAL TUBE DEFECTS	TRANSFER OF AFP FROM FETAL CSF
ESOPHAGEAL ATRESIA	REDUCED TURNOVER, LACK OF SWALLOWING
CONGENITAL NEPHROSIS	LOSS OF FETAL AFP FROM RENAL DEFECT
MULTIPLE PREGNANCY	INCREASED PRODUCTION
Rh INCOMPTABILITY	? INCREASED PRODUCTION

TABLE III. POSSIBLE FUNCTIONS OF AFP DURING FETAL DEVELOPMENT

OBSERVATION	POSSIBLE FUNCTION
STRUCTURE HOMOLOGY WITH ALBUMIN, INVERSE SERUM CONC. TO ALBUMIN	FETAL ALBUMIN, CARRIER AND OSMOTIC
IMMUNOSUPPRESSIVE	BLOCKS MATERNAL REJECTION OF FETUS
BINDS ESTROGEN	PROTECTS FETUS FROM MATERNAL ESTROGEN
PRODUCED DURING LIVER LOBULE FORMATION	DEVELOPMENTAL ORGANIZATION SIGNAL

TABLE IV. EVIDENCE FOR IMMUNOSUPPRESSION BY AFP

1. AFP IS HIGH IN FETAL SERUM AND MATERNAL SERUM
2. AFP IS HIGH IN SOME CANCER PATIENTS
3. AFP IS HIGH IN PATIENTS WITH ATAXIA TELANGIECTASIA
4. AFP INHIBITS MITOGEN-INDUCED BLAST TRANSFORMATION
5. AFP INHIBITS MIXED LYMPHOCYTE REACTIONS (I REGION)
6. AFP INHIBITS GENERATION OF PFC *IN VITRO*
7. AFP INHIBITS GENERATION OF CYTOTOXIC T CELLS
8. AFP ACTIVATES SUPPRESSOR CELLS
9. AFP IS MADE BY T CELLS (?)

TABLE V. EVIDENCE AGAINST A SIGNIFICANT IMMUNOSUPPRESSIVE ROLE FOR AFP

1. NO EFFECT IN VIVO (HEPATOMA, TYROSINEMIA, ATAXIA TELANGIECTASIA, PREGNANCY)

2. PASSIVE TRANSFER OF AFP NOT EFFECTIVE IN VIVO

3. FETUS CAN RESPOND TO IMMUNIZATION IN UTERO

4. AFP DOES NOT CROSS THE PLACENTA IN RUMINANTS

5. IN VITRO RESULTS NOT REPRODUCIBLE IN DIFFERENT LABORATORIES

6. MANY REDUCTIONS OF IN VITRO RESPONSES NOT BIOLOGICALLY SIGNIFICANT

7. ONLY SOME ISOLATED AFP PREPARATIONS ACTIVE IN VITRO

8. ISOLATED ALBUMIN OFTEN HAS SIMILAR EFFECT IN VITRO

9. NO GREATER EFFECT OF HEPATOMA SERA (10,000 µg/ml) OVER NORMAL SERA (.03 µg/ml)

10. FETAL BOVINE SERUM (USED FOR STUDIES ON IN VITRO IMMUNE RESPONSE) CONTAINS HIGH CONCENTRATIONS OF AFP (1,000 µg/ml)

11. BIRDS (EGG WHITE) CONTAIN AFP EQUIVALENT)

TABLE VI. AFP PRODUCTION BY NORMAL FETAL AND ADULT TUMOR TISSUE

FETAL TISSUE	AFP (% of total)	ADULT TUMOR	AFP (% of patients > 40 ng/ml)
YOLK SAC	61	ENDO. SINUS	100
LIVER	30	HEPATOMA	80
BILIARY TRACT	?	BILIARY TRACT	25
STOMACH	3	GASTRIC	15
COLON	3	COLON	3
SMALL INTESTINE	2	---	-
LUNG	<1	PULMONARY CARCINOMA	7
BREAST	<1	ADENOCARCINOMA	2
OTHER	<1	SARCOMA, ETC.	<1

TABLE VII. LEVELS OF EXPRESSION OF ONCODEVELOPMENTAL GENE PRODUCTS BY TUMORS

PRODUCT	FIRST LEVEL Fetal or adult tissue which normally produces it	SECOND LEVEL Closely related embryologically (same cell line)	THIRD LEVEL More distantly related (same cell line)	FOURTH LEVEL Different cell line
AFP	ENDODERMAL SINUS TUMOR, HEPATOMA	PANCREAS, GI, BILIARY TRACT	PULMONARY BREAST	SARCOMA
CEA	COLON	PANCREAS, GASTRIC LIVER	PULMONARY BREAST	SARCOMA
HORMONE (eg ACTH)	PITUITARY ADENOMA	MEDULLARY CA OF THYROID	OAT CELL LUNG	PULMONARY ADENOCA

TABLE VIII. PROPERTIES OF RAT HEPATOCELLULAR CARCINOMAS ASSOCIATED WITH HIGH AFP PRODUCTION

RAPID GROWTH RATE

ABNORMAL CHROMOSOME NUMBER

DEGREE OF HISTOLOGIC DIFFERENTIATION

CARCINOGEN USED

TABLE IX. PATTERNS OF AFP PRODUCTION AND CELLULAR CHANGES DURING HEPATOCELLULAR CARCINOGENESIS IN RATS

PATTERN	CARCINOGENS	DEARLY	LATE	HEPATOMAS
I	DEN (Low Dose)	NONE NO AFP ↑	NODULES AND OVAL CELLS, RAPID AFP ↑	MULTIPLE ALL AFP+
II	DAB	NECROSIS AND PROLIFERATION, OVAL CELLS, SLIGHT TO MODERATE AFP ↑	NODULES AND OVAL CELLS, RAPID AFP ↑	MULTIPLE ALL AFP+
III	FAA ETHIONINE	NO MORPHOLOGIC CHANGE, MODERATE PROLONGED AFP ↑	NODULES AND OVAL CELLS, NEAR NORMAL AFP	AFP+ AND AFP-
IV	WY 14643 AZASERINE	HEPATOCYTE PROLIFERATION SLIGHT TEMPORARY AFP↑	NODULES ONLY NEAR NORMAL AFP	ALL AFP-

ALPHA-FETOPROTEIN AND THE IMMUNE RESPONSE

T. B. Tomasi, Jr.

Department of Immunology, Mayo Clinic, Rochester, Minnesota 55901

Our interest in alpha-fetoprotein (AFP) as a possible immunoregulatory factor arose out of experiments designed to explain the delayed development of IgA plasma cells in the GI tract of newborn mice. We postulated that a circulating immunosuppressive factor might be responsible for the general immunological immaturity seen in the mouse (and certain other species) during fetal life and for several weeks following birth. Our first experiments involved injection of mouse amniotic fluid (MAF) from birth until young adulthood (5-6 weeks) in order to continue to supply the presumptive inhibitor. The results of these experiments showed that the *in vivo* administration of MAF inhibited the development of the plaque forming response (PFC) to sheep red blood cells (SRBC), particularly in the IgG and IgA classes.[1] Subsequently, our laboratory and several others[2] have shown that MAF does indeed inhibit the antibody response to T dependent antigens as well as a variety of cell-mediated reactions in which T cells are involved. Evidence has been presented that several T independent responses such as the antibody response to DNP-Ficoll and DNP-POL are not inhibited by MAF[3] although the LPS response (also T independent) is very sensitive to inhibition by a non-dialyzable factor(s) (?AFP) present in MAF.

Initially, fractionation studies of MAF showed that the majority of the inhibitory effect was present in isolated and apparently highly purified preparations of AFP. Subsequent work[4], however, has presented evidence for a second, non-AFP-suppressive factor present in MAF and newborn serum. The non-AFP immune inhibitor, which is currently under active investigation, has not yet been isolated in homogenous form. Although thus far the data suggests the existence of two suppressive factors in these fluids, it has not been completely excluded that a single potent suppressive substance exists which contaminates both AFP and non-AFP preparations in amounts which have escaped detection by the methods employed.

There is little disagreement among investigators that amniotic fluid and fetal and newborn sera contain a potent suppressive factor or factors. However, it should be mentioned that certain workers have found variable suppressive effects of AFP depending on the system tested and in some cases even enhancement has been noted.[2] Although the reasons for these discrepancies have not

---

This work was supported by grant CA-l8204 and by National Institutes of Health Grant HD-09720.

been completely explained, it seems likely that a variety of technical problems inherent in both the methods of isolation of AFP as well as differences in *in vitro* test systems employed to study suppression explain these discrepancies. For example, many of the techniques used to isolate AFP may inactivate it either because of its inherent structural lability or the necessity for association with low molecular weight factors. The source of AFP is important since it has been reported that only certain types of AFP (the most negatively charged species) are suppressive[5,6] and the proportion of suppressive to nonsuppressive variants of AFP differ in various fluids.[7] In addition, we have found that culture conditions (such as cell density and type and concentration of supporting sera) are extremely important in analyzing for inhibitory properties of the various suppressive factors including AFP.

Several hypotheses have been put forth to explain suppression by MAF and AFP. These include inhibition of the T helper and/or amplifier cells and the development of suppressor T cells. The suppressor cell hypothesis is especially attractive since newborn animals have relatively high levels of T suppressor cells and data presented from Wigzell's laboratory[8,9] suggest that AFP may affect an Ly 1 positive T cell which in turn induces Ly 2,3 suppressor cells via an Ly 1,2,3 cell in the "feedback" suppressive loop as described by Cantor et al.[10] Our recent data outlined below suggest an alternative possibility in which the neonatal suppressive factor has an effect at the macrophage level.

METHODS

CBA/J and C3H/Hej mice were purchased from Jackson Laboratories, Bar Harbor, Maine. Pregnant Ha/ICR or CBA mice were used as a source of amniotic fluid (MAF) and were bred locally at the Mayo Clinic. The recombinant strains used for the MLR studies were produced locally in our mouse colony by Dr. Chella David.

<u>Preparation of MAF, AFP and MAF-AFP</u>: MAF was collected from pregnant mice as previously described,[11] diafiltered using an Amicon P-10 membrane and flushed with .14 M phosphate buffered saline at pH 7.2. Diafiltering was repeated 4-5 times until over 90% of the dialyzed material in the original fluid was removed and then reconstituted to 2 OD units measured at 280 nm. In some experiments MAF was obtained sterilely from the amniotic sac and used without any further treatment.

AFP was isolated as previously reported.[4] Briefly, the dialyzed MAF was applied to an anti-normal mouse serum (NMS) affinity column. The PBS effluent was tested by gel diffusion against anti-AFP and anti-normal mouse serum. All AFP positive fractions were pooled and concentrated and reapplied to the column if they showed any reaction to the anti-NMS on gel diffusion. Purity of the AFP was also examined by polyacrylamide gel electrophoresis, and by Ouchterlony gel diffusion and immunoelectrophoresis using anti-NMS and anti-MAF. We have employed six different schemes for isolating AFP, including some of the methods used by other workers[8,12] and found that the simple one-step method outlined above yields preparations as homogeneous as those using some of the more complicated techniques. Moreover, the AFP prepared by this method retains its immunosuppressive properties and is obtained in high yield. The key to this method is to employ a carefully selected anti-normal mouse serum which will remove the serum components in MAF. MAF-AFP was prepared by affinity chromatography using an anti-AFP antisera. Specific antibody from this antisera was isolated on specific antigen columns (using preparations containing 90+% AFP) and the eluted gammaglobulin coupled to Sepharose 4B.

T dependent lymph node proliferative assay: This assay was performed with minor modifications essentially as described by Alkans.[13] Briefly, antigen was emulsified in H37ra complete adjuvant (Difco Laboratories, Detroit, Michigan) and was injected into the CBA mice at the base of the tail. Eight days later the draining lymph nodes of the inguinal and para-aortic regions were removed, cells teased apart and passed through a steel mesh screen. Cells were washed four to five times and the cell numbers adjusted to 4×10^5/well in culture media. Culture media consisted of RPMI 1640 supplemented with 5% heat inactivated horse serum (GIBCO, Grand Island, New York), 10 mM HEPES buffer, 4×10^{-5} M 2 mercaptoethanol, penicillin, streptomycin and glutamine. The antigens were added in varying concentrations (250 µg/ml ovalbumin or OVA and 50 µg/ml PPD) and the proliferative response measured after four days by ^{3}H-thymidine incorporation.

Peritoneal exudate cells were obtained from nonimmunized syngeneic mice by flushing the peritoneal cavity with 5-6 ml of cold Hank's buffered salt solution or .35 M sucrose. Antigen pulsing experiments were performed by either pulsing the PECs directly with antigen for two hours and then washing the cells for 4-5 times to remove free OVA; alternatively, the PECs were allowed to adhere to plastic, the nonadherent cells removed by washing and the adherent PECs then pulsed for two hours, washed and the sensitized lymph node cells added.

Mixed lymphocyte reactions: The primary mixed lymphocyte reactions were performed as previously described.[14] Reactions were carried out using $.5 \times 10^6$ live cells/well aliquoted into flat-bottomed Falcon plastic microtiter plates. 1×10^6/well live mitomycin treated (50 µg/ml) or irradiated (3000 R) cells were added as stimulators. 20 µl of the various MAF preparations (whole MAF, AFP, AFP-MAF) were added to the culture well to give a final concentration of 200 µg/ml and after three days of incubation, the proliferative response was measured using ^{3}H thymidine incorporation. Incorporation is expressed as counts per minute ± the standard deviation. At least three experiments were performed with each of the strain combinations and five different MAF, AFP and MAF-AFP preparations were employed.

RESULTS

Experiments outlined in Figure 1 suggest that MAF inhibits antigen induced T cell proliferation via an effect on an adherent antigen presenting cell probably a cell of the monocyte-macrophage lineage. In this experiment regional lymph cells were obtained from an animal previously sensitized to ovalbumin and PPD and the proliferative response measured *in vitro* after the addition of antigen, either added directly to the culture or pulsed on peritoneal exudate cells (PEC). Evidence has been previously presented that this proliferative assay is T cell dependent (inhibited by anti-θ plus complement) and requires the presence of macrophages.[13] For example, the proliferative response of nylon column purified (macrophage depleted) sensitized T lymph node cells cultured in the presence of OVA is markedly decreased compared to the unfractionated lymph node cells. The addition of OVA pulsed peritoneal exudate cells (PEC) restores the proliferative capacity of purified T cells (Figure 1C). The data in this figure indicates that when PECs were cultured for 24 hours in the presence of MAF, pulsed with OVA, washed and added to sensitized T cells, that the proliferative response was suppressed (Figure 1B) compared to PEC incubated in the absence of MAF or controls in which MAF was added only during the last hour of the 24-hour incubation period. This suggests that MAF interferes with antigen presentation. Other experiments (data not shown) indicated that suppression is not due to inhibition of antigen uptake by the macrophage. The level at which MAF interferes with antigen presentation is unknown but some preliminary data suggest that perhaps MAF interferes with the T-macrophage interactions which require I region compatibility between the interacting cell populations. Experiments (K. Suzuki and T. Tomasi, submitted for publication) in which spleen cell preparations are partially depleted of macrophages demonstrated that their response to PHA is not longer inhibitable by MAF and adding back PECs restores suppression by MAF. This again emphasizes the necessity of having macrophages in the preparation for suppression by MAF.

The question of whether MAF and AFP are "nonspecific" inhibitors of proliferation of cells was approached by studying suppression of the mixed lymphocyte reaction (MLR) by these factors. As summarized in Table 1, various MHC region induced MLRs are suppressible by MAF and AFP. However, proliferative reactions induced by certain Mls loci and non-MHC loci are either not inhibited or actually stimulated by the same MAF and AFP preparations that produced inhibition of the MHC induced MLRs. This data suggests that suppression is not simply a result of nonspecific inhibitors of proliferative reactions regardless of the cell types involved. The biological significance of the genetic restrictions in suppression is unknown.

While there is little doubt that MAF and AFP suppress a variety of immune reactions *in vitro*, a key question is whether similar effects apply *in vivo*.

In order to examine the *in vivo* effect of MAF, the following experiments were performed. Adult (8-12 weeks) CBA mice were given sterile LPS negative (by limulus assay) MAF intravenously for 3 days and then injected with varying doses of SRBC. MAF administration was continued daily for 4 days following SRBC administration and then on day 4 direct splenic PFCs were measured. Blood levels of AFP were measured by radioimmunoassay at the time of plaquing. The results are presented in Table 2. This data is a summary of 5 different experiments. There is a statistically significant inhibition of the PFC response which is particularly prominent in the 1×10^7 SRBC group. Inhibition by MAF is significant but less in the optimally and supraoptimally immunized groups. Mean levels of AFP achieved at the time of plaquing were in the range of 200,000 ng/ml which is about 1/10 that found in the sera of newborn animals.

Studies have also been performed[15] on the effect of the administration of MAF *in vivo* on tumor growth. Note in Table 3 that pretreatment of mice with MAF increases the rate of growth and the number of tumors in histo-incompatible animals while Table 4 illustrates that such treatment increases the mortality in histocompatible tumor bearing animals. Also the number of tumor "takes" when syngeneic animals are challenged with small numbers of histocompatible tumor cells is increased by MAF as shown in Table 5. In these studies preliminary experiments were performed to determine the number of tumor cells which should be injected to give approximately 25-50% tumors in the control group (NMS diluted to the same concentration as MAF).

Table 6 presents data on the effect of *in vivo* treatment of adult mice on the development of tolerance to human gamma globulin (HGG). Six week old A/J mice were tolerized with deaggregated HGG (DHGG) and then given antigen (aggregated HGG or DHGG) 15 days later according to the protocol described by Chiller and Weigle.[16] Ten days after antigen administration, splenic PFC were performed using goat cells coated with HGG. It should be noted in Table 6 that suppression in the MAF + DHGG group is significant in the IgG and especially IgA classes and that suppression is greater than the sum of the inhibition expected by either MAF or DHGG alone (51% and 88% for IgG and IgA respectively). It is emphasized that similar to the work outlined above with the SRBC response and tumor growth that whole MAF was administered and not AFP. Thus, this data does not directly demonstrate that AFP is involved in the suppressive effects since other substances in MAF rather than or in addition to AFP could be responsible. Work is now in progress investigating the effects of the *in vivo* administration of AFP on tumor growth and the ease of induction of tolerance to allogeneic cells as well as soluble antigen such as HGG.

DISCUSSION

The evidence presented in this paper indicates that MAF inhibits the proliferative responses of T cell induced by both soluble antigens and alloantigens. Studies previously reported from our laboratory[4] suggest but do not prove that two inhibitory factors may be present in MAF: AFP and a non-AFP suppressor which is detected in samples of MAF depleted of AFP by affinity chromatography. However, whatever the nature of the suppressive factor(s) in MAF, it appears to show some specificity as indicated by the failure to inhibit non-MHC induced proliferative reactions. The nature of the genetic restrictions in the MLR are of considerable interest and are now being actively pursued in several laboratories. Similar genetic restrictions have been reported by Peck et al.[17] although our data is not in agreement with theirs in certain details which are discussed at length in a previous publication (C. Krco, submitted for publication).

In regard to the mechanism of inhibition by MAF, we have obtained evidence that it does not function at least in the *in vitro* lymph node proliferative assay and in the splenic response to SRBC via the induction of an I-J positive, cytoxan sensitive, suppressor T cell as suggested by Murgita et al.[8] Rather, we believe the data point to inhibition at the macrophage level, although the exact mechanism has not been established. It is possible that MAF inhibits the Ia positive antigen presenting cell directly or perhaps via an effect on T-macrophage interactions. However, it is also possible that MAF may induce another cell type in PECs, possibly an Ia macrophage, to suppress the antigen presenting cell. Further studies are necessary to clarify these points. In addition, the mechanism of suppression by MAF administered *in vivo* could theoretically be different than that found under the artificial conditions employed *in vitro*.

Despite the fact that MAF and, in many laboratories, AFP have been reported to be suppressive to a variety of *in vitro* tests, the question still remains as to the biological significance of these findings. Experiments presented here suggest that the administration of MAF will, indeed, suppress the animal's response to SRBC. In these experiments, serum AFP levels were measured by radioimmunoassay in order to monitor the levels of suppressive factors achieved relative to those seen in the pregnant and neonatal animals. The levels achieved by our regime were close to those seen in normal pregnancy (or slightly higher) but were in general about 1/10 those found in the fetus and newborn animal. With these doses of MAF, we were able to achieve significant suppression, particularly when suboptimal doses of antigen (SRBC) were given. It is noteworthy that supraoptimal suppression was less suppressed in these *in vivo* experiments. One might have expected that if MAF induced suppressor cells, that the degree of suppression with supraoptimal SRBC would have been even greater since high doses of antigen promote suppressor cell development. We were unable, with the amount of MAF administered in these experiments, to achieve the goal of reproducing the levels of AFP seen in the newborn animal. We are now attempting by concentration of MAF to administer larger amounts in order to achieve newborn levels and to determine if these adult animals then behave immunologically as do newborns. In the preliminary experiments present here, there is evidence that tolerance to HGG is more readily induced in animals treated with MAF but in these studies only low levels of AFP (in the 5-8,000 ng/ml range) were achieved and statistically significant enhancement of tolerance was seen only in the IgA class in four different experiments similar to that presented in Table 6. Thus, further work on tolerance, both to soluble antigens as well as to cells, is indicated. Similarly, as reported here and elsewhere,[15] there is an indication that MAF treated animals are more susceptible to tumors. In this study the number of tumor cells required to initiate a tumor is reduced and the growth rate of established tumor is accelerated in the MAF treated animals despite the fact that we achieved only low levels of AFP (5000 ng/ml). It should be mentioned that although we monitored AFP levels as an indication of the amounts of the inhibitor administered and the closeness with which we approached the pregnant and neonatal states, these *in vivo* experiments in no way prove that AFP is involved in suppression. It may well be that another inhibitor, acting by itself or in concert with AFP, is responsible for suppression. It will be necessary in subsequent experiments to use highly purified preparations of AFP. A very recent study[18] does suggest that AFP per se promotes the growth, delays regression, increases mortality, and lowers the threshold dose of virus necessary to induce tumors in the Maloney sarcoma virus system.

Although indirect, evidence suggesting that AFP may be involved in suppression *in vivo* has also been obtained in studies of cancer-prone families and patients with chronic active hepatitis and lymphomas. These studies which have been previously reviewed[19] suggest that immune suppression as determined by the PHA response and depression of the MLR is better correlated with the presence of cells bearing AFP on their surface than with circulating AFP levels. This appears to be a reversible suppression in both the cancer-prone patients and in chronic hepatitis since culture of the peripheral blood cells for 7 days followed by washing results in a return of their PHA and MLR reactivity. Supernates of these 7 day cultures are suppressive to normal lymphocytes. The majority of patients with cell surface AFP had normal serum AFP levels. Thus, this evidence suggests the possibility that either cell bound AFP or a factor correlating with the presence of AFP on the surface of cells may be responsible for suppression in certain human disorders. It should be emphasized that we have not been able to demonstrate that a significant proportion of patients with sporadic forms of cancer have circulating cells with AFP as determined by immunofluorescence. Moreover, several cancer-prone kindreds have been studied in which AFP has not been found on cell surfaces. A sensitive (and more subjective) technique such as a cellular radioassay should be employed to test whether the negative cases may be due to the relative insensitivity of the immunofluorescent procedure in detecting cell bound AFP.

REFERENCES

1. Ogra, S.S., Murgita, R.A., Tomasi, T.B., Jr.: Immunol. Commun. 3:497, 1974.
2. Tomasi, T.B., Jr.: Cell. Immunol. 37:459, 1978.
3. Murgita, R.A., Wigzell, H.: Scand. J. Immunol. 5:1215, 1976.
4. Labib, R., Tomasi, T.: Immunol. Commun. 7:223, 1978.
5. Zimmerman, E.F., Voortung-Hawking, M., Michael, J.G.: Nature (London) 265:354, 1977.
6. Lester, E.P., Miller, J.B., Yachnin, S.: Proc. Nat. Acad. Sci. USA 73:4645, 1976.
7. Yachnin, S., Lester, E.: Clin. Exp. Immunol. in press.
8. Murgita, R.A., Goidl, E.A., Kontiainen, S., Wigzell, H.: Nature (London) 267:257, 1977.
9. Murgita, R.A., Goidal, E.A., Kontiainen, S., Beverley, P.C.L., Wigzell, H.: Proc. Nat. Acad. Sci. USA 75:2897, 1978.
10. Cantor, H., McVay-Boudreau, L., Hugenberger, J., Naidorf, K., Shen, F.W., Gershon, R.K.: J. Exp. Med. 147:1116, 1978.
11. Murgita, R.A., Tomasi, T.B., Jr.: J. Exp. Med. 141:269, 1975.

12. Sheppard, H.W., Sell, S., Trefts, P., Baku, R.: J. Immunol. 19:91, 1977.
13. Alkan, S.S.: Eur. J. Immunol. 8:11, 1978.
14. Murgita, R.A., Tomasi, T.B., Jr.: J. Exp. Med. 141:440, 1975.
15. Tomasi, T.B.,Jr., Dattwyler, R.J., Murgita, R.A., Keller, R.H.: Transactions of the Association of American Physicians 88:293, 1975.
16. Chiller, J.M., Weigle, W.O.: IN Contemporary Topics in Immunobiology, Hanna, M.D., Ed., Vol. 1, Plenum Press, New York, P. 19, 1972.
17. Peck, A.B., Murgita, R.A., Wigzell, H.: J. Exp. Med. 147:667, 1978.
18. Gershwin, M.E., Castles, J.J., Makishima, R.: J. Immunol. 121:2292, 1978.
19. Tomasi, T.B., Jr.: Ann. Rev. Med. 28:453, 1977.

TABLE 1. EFFECT OF MAF AND AFP ON THE MIXED LYMPHOCYTE REACTION (MLR) SPECIFIC FOR VARIOUS HISTOCOMPATIBILITY LOCI

Histocompatibility Differences Stimulating MLR	Inhibition of MLR by MAF and AFP
MHC*	+
I region	+
K region	+
D region	+
MLS locus	- or +
Non MHC	0 - (stimul.)

*MHC = major histocompatibility complex.

Summary of data from Krco, C., Johnson, E., David, C.S., and Tomasi, T.B., submitted for publication.

TABLE 2. IN VIVO SUPPRESSION OF SRBC PFC RESPONSE BY MAF

Dose SRBC	% Suppression MAF	Serum AFP ng/ml
Suboptimal (1×10^7)	84%	190,200
Optimal (5×10^8)	49%	168,000
Supraoptimal (1×10^{10})	45%	217,500

1 cc of MAF (dialyzed-adjusted to 2 OD units) administered for 3 days before and 4 days after SRBC injection. PFC on day 4. Mean of 5 experiments with different pools of MAF. Four mice in each experiment. Percent suppression compared with PBC injected control.

TABLE 3. EFFECT OF MAF ON GROWTH OF SARCOMA I IN HISTOINCOMPATIBLE MICE

	No. Animals	No. with Tumors	Mean Wt. Tumor (mg)
Controls	23	9	1.3
MAF	24	18	6.0

C57/Bl mice sacrificed 13 days after injection of 10×10^6 sarcoma I cells SC. Tumors dissected and weighed. Differences in weight significant ($P < .05$).

TABLE 4. EFFECT OF DAILY TREATMENT OF MICE WITH AMNIOTIC FLUID ON NUMBER OF TUMORS DEVELOPING IN MICE HISTOCOMPATIBLE AT H2 WITH TUMOR*

Treatment	Number of Animals	Number with Tumors			Number Dead d60
		d20	d35	d60	
Normal Mouse Serum	12	2	3	3	2
Amniotic Fluid	12	6	10	10	9

*B6AF1/J mice injected on alternate days with 1 ml or 0.5 ml normal mouse serum or amniotic fluid. Mice injected with 10^4 viable sarcoma I cells subcutaneously on day 3.

TABLE 5. EFFECT OF DAILY ADMINISTRATION OF AMNIOTIC FLUID ON NUMBER OF TUMORS DEVELOPING IN MICE GIVEN SMALL NUMBERS OF TUMOR CELLS*

Mice and Tumors Histocompatible at H-2 Locus

Injected with	No. of Mice	No. with Tumors		
		d12	d16	d24
Normal Mouse Serum	12	1	3	3
Amniotic Fluid	12	6	9	10
			$P<.01$	

*DBA/2 mice injected on alternate days with 1.0 and 0.5 ml of NMS or amniotic fluid. Given 100 viable mastocytoma cells on day 3.

TABLE 6. TOLERANCE TO HUMAN GAMMA GLOBULIN

Treatment	No. Mice	PFC			Serum AFP ng/ml
		IgM	IgA	IgG	
PBC	11	8 ± 2	85 ± 20	876 ± 190	205
MAF	12	2 ± 1	66 ± 12	636 ± 101	5598
DHGG	12	4 ± 1	47 ± 15	675 ± 140	801
DHGG + MAF	12	2 ± 1	8 ± 2	232 ± 58	5755

100 µg deaggregated human gamma globulin (DHGG) injected into 6 week A/J mice at day 0. Heat aggregated HGG injected on day 15 and 25 and PFC (goat RBC coated with HGG) performed 5 days later. MAF (0.5 ml) given daily from day 0 to day 7.

Figure 1.

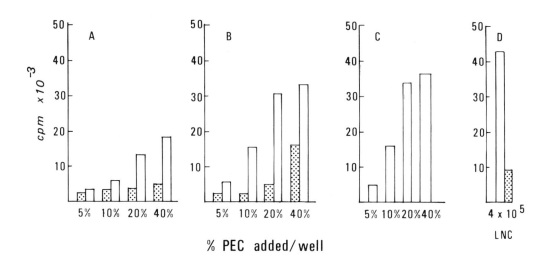

% PEC added/well

A: Various numbers of adherent PECs from non-immunized CBA/J mice were pulsed with OVA (250 μg/ml) in the microtiter plate, washed and then cultured in the presence of MAF (200 μg/ml) either for 24 hours (dotted columns) or, as a control of carry-over, the last hour of the 24 hour preculture period (open column). The PECs were then washed and 4×10^5 OVA-primed T lymph node cells added.

B: Adherent PECs were cultured for 24 hours in the presence of 200 μg/ml of MAF (dotted column), and then pulsed with OVA (250 μg/ml) for 2 hours. After washing, 4×10^5 OVA-primed T lymph node cells were added. As a control (open column), MAF (200 μg/ml) was added one hour before the T lymph node cells were added.

C: Adherent PECs were pulsed with OVA (250 μg/ml) for 2 hours, washed, and then 4×10^5 OVA-primed T lymph node cells were added.

D: 4×10^5 OVA-primed lymph node cells were cultured with OVA (250 μg/ml) for 4 days in the presence (dotted column) or absence (open column) of MAF.

Data from K. Suzuki and T. B. Tomasi, submitted for publication.

THE USE OF CEA AS A TUMOUR MARKER

J. Shuster, M.D., Ph.D.[1] and P. Gold, O.C., M.D., Ph.D., F.R.S.(C)[2]

[1]*Clinical Research Association of the National Cancer Institute of Canada*, and [2]*Associate of the Medical Research Council of Canada; McGill Cancer Centre, McIntyre Medical Sciences Building, 3655 Drummond Street, Montreal, Quebec, Canada H3G 2Y6*

Biochemical parameters that correlate with the presence of malignancy, the location of the tumour and the size of the tumour burden are obviously of great importance to the clinical oncologist. The studies of Foley just over one-quarter of a century ago indicated that cancer cells produce unique substances which may elicit a host anti-tumour immune response (1). This initial demonstration of unique tumour antigens was made possible by the development of inbred strains of mice. He showed that tumour grafts, transplanted between syngeneic (genetically identical) mice were rejected by the recipient of the transplanted tumour tissue. Since one was dealing with tissues transplanted between otherwise genetically identically identical mice, the only means by which the tumour cells could be rejected was through the development, during the process of malignant transformation, of unique antigens on the tumour cell surface. Since the method used to demonstrate the presence of tumour antigens was the rejection of transplanted tumour tissue, these intially demonstrated antigens were called tumour specific transplantation antigens (TSTA). Indirectly, similar conclusions were reached with respect to human tumours when it was demonstrated that the tumour bearing host could produce a specific humoral and/or cellular antitumour immune response against his own tumour (2-4). These studies sparked considerable investigations into the nature of these unique tumour components.

CARCINOEMBRYONIC ANTIGEN

The role of tumour markers in the management of patients with cancer is best illustrated in the extensive studies related to the carcinoembryonic antigen (CEA). CEA was identified just over a decade ago employing an approach involving the immunization of xenogeneic animals with extracts of adenocarcinomas of the human colon. The heterologous antitumour antisera prepared in these rabbits were rendered tumour specific by absorbing the antisera with an excess of the corresponding normal colonic tissue in order to removal all the antinormal tissue antibody activity from the antiserum produced as described above. In agar gel, the absorbed antiserum detected a component unique to malignancies of entodermally derived digestive system epithelium (5,6). The antigen was apparently absent from primary cancers of all other tissues but was detected, by precipitin reactions in agar gel, in extracts of the embryonic and fetal gut, pancreas and liver in the first two trimesters of gestation. Because of the distribution of the antigenic component(s) so detected in both malignant and embryonic tissue, this molecule was termed the carcinoembryonic antigen (CEA) of the digestive system. Employing a variety of immunocytologic techniques, CEA was localized to the tumour cell surface, more precisely to the glycocalyx immediately adjacent to the surface membrane (7). Antibodies specific for CEA are capable of inducing a polar redistribution or capping of CEA on the tumour cell surface, suggesting that CEA is a peripheral membrane glycoprotein that is part of the fluid mosaic of the cell membrane (8). As will be seen in the subsequent clinical studies, from this site on the cell surface, the CEA is apparently released to the surrounding body fluids from which it will eventually appear to the circulation where it can be measured by sensitive radioimmunoassay procedures.

PHYSICO-CHEMICAL PROPERTIES OF CEA

CEA was initially isolated from hepatic metastases of adenocarcinoma of the colon following perchloric acid extraction, and a combination of molecular sieve chromatography and block electrophoresis. This molecule is a large glycoprotein with molecular weight of approximately 200,000. Highly purified preparations of CEA manifest a considerable degree of both intermolecular and intramolecular heterogeneity (9). The carbohydrate to protein ratio in individual CEA preparations varies considerably but most often the molecule consists of about 60% carbohydrate and 40% protein. Aspartic acid or asparagine and glutamic acid or glutamine, threonine and serine constitute the major amino acid residues, while N-acetyl-glucosamine represents the major monosaccharide moiety of this molecule (10). The heterogenous electrophoretic behavior of CEA is in part attributable to the variable content of sialic acid residues in individual CEA preparations and individual CEA molecules. CEA is a cruller-shaped particle of single chain structure in which multiple intrachain disulphide bonds maintain the overall conformation of these molecules.

IMMUNOCHEMISTRY OF CEA

Although the initial studies employing the relatively insensitive method of precipitation reaction in agar gel to assess the tissue distribution of CEA suggested the tumour specificity of the CEA-anti CEA reaction, more recent studies, particularly the clinical studies relating to the radioimmunoassay for CEA, have indicated that the product measured by radioimmunoassay in the circulation is not specifically related to malignant transformation (12). The so-called "false positive" results in the CEA radioimmunoassay is the result of either CEA, or other "CEA-like" components that are present in non-enteric neoplasms and perhaps various normal tissues. Within the past several years, twelve cross-reacting antigens have been identified using the antisera prepared against purified CEA (9,12). Six of these twelve cross-reacting antigens appeared to be immunologically identical but the relationship of the remaining antigens to each other requires further study. Another problem that may complicate the demonstration of the tumour specificity of CEA is that true monospecificity of the anti-CEA antisera cannot be assured unless extensive absorption with normal tissue extracts has been performed. If an antiserum is used which is directed principally against the common antigenic sites of two different macromolecules, one of which is tumour specific while the other is not, complete identity of the two substances will be erroneously concluded, and the apparent detection of the tumour material or normal tissues will obscure its tumour specificity. Although absorption of antisera can increase specificity of the immune reaction, this method may be limited practically by the choice and availability of suitable quantities of material(s) for absorption. After extensive absorption, the titer of truly monospecific antibodies against a single tumour-specific antigen determinant may be quite low. A potential approach to overcome these difficulties lies in a better understanding of the molecule structures of CEA and other "CEA-like" molecules. To this end, Drs. Krantz and Ariel, in our laboratories, have undertaken a study of the structure of CEA by preparing a group of proteolytic fragments which vary in size from 5 to 20% of the intact CEA molecule (13). Degradation of CEA requires prior denaturation by boiling in SDS-mercaptoethanol followed by limited proteolytic digestion with protease V8, trypsin or chymotrypsin. Each enzyme produced a unique pattern on SDS-PAGE of fragments ranging in molecular weight from 10,000 to 40,000, in addition to dispersed material in the size range of 45,000 to 70,000. A given enzyme produces identical low molecular products when different CEA preparations are employed and compared. Using protease V8, 16 distinct fragments are obtained, while 9 are seen with chymotrypsin and 9 with trypsin. A number of these fragments have been eluted from the gels and their ability to inhibit the standard radioimmunoassay for CEA has been performed. Of the 12 fragments tested, only one failed to inhibit the CEA radioimmunoassay. Four of these fragments appeared to reach a plateau at much less than 100% inhibition indicating the lack of some of the CEA antigenic groupings that are detected in the radioimmunoassay. Of interest was the fact that the smallest protease V8 fragment, of about 14,000 in molecular weight, was one of the most potent inhibitors of all the fragments tested. At relatively low concentrations, this fragment inhibited the CEA-anti-CEA reaction to about 70% before plateauing.

Several interesting conclusions can be derived from this study. Most of the fragments so prepared had relatively low carbohydrate content, indicating that the CEA immunoreactivity is determined, to a considerable extent at least, by the protein content of the molecule. Furthermore, since CEA preparations from different tumours yielded identical low molecular weight fragments, there must be a substantial degree of sequence homology between CEA preparations. In addition, the inhibition studies performed with the fragments eluted from the gels indicate that the putative tumour-specific sites may comprise a relatively small portion of the CEA molecule. Nearly all the fragments demonstrated substantial immunoreactivity implying that a considerable degree of cross-reactivity of individual fragments must exist. This would occur if the majority of fragments are derived from the same limited region of the CEA molecule or, alternatively, if common antigenic groupings are represented in a repetitive fashion over a large portion of the CEA molecule. Further biochemical and immunological analysis of the CEA fragments should resolve these questions. One or more of these fragments may be capable of producing antisera that will increase the specificity of the CEA radioimmunoassay.

RADIOIMMUNOASSAY FOR CEA

Over the past decade, a number of radioimmunoassays have been developed for CEA. These have resulted in a burgeoning literature on the usefulness of CEA as a tumour marker. Although these different assays have been performed in different fashions and with different antigen and antibody preparations, similar conclusions may be reached from these distinct CEA radioimmunoassay studies. In the interpretation of the clinical data obtained, the smoking status of the individual must be considered. For example, using the commercially available Z-gel radioimmunoassay for CEA (Hoffman-LaRoche), about 20% of all smokers without neoplasms have low but persistently elevated CEA levels, (between 2.5 and 5 ng/ml). In contrast to normal non-smokers where CEA levels are invariably below 2.5 ng, the normal level for smokers should thus be considered at 5 ng/ml.

The clinical data indicates that the highest incidence of elevated CEA levels (80%) are found in tumours of the colon and rectum (14-16). The incidence of CEA elevations vary with the stage of the tumour. Thus, in Dukes stage A colorectal neoplasms, the lowest incidence of CEA positivity (45%) is observed. In Dukes B, C and D lesions, the incidence rises to 55%, 70% and 90% respectively. In addition, CEA elevations are seen in other enteric neoplasms such as adenocarcinomas of the stomach and pancreas.

However, despite these observations, the currently employed CEA radioimmunoassays deviate from the ideal in that they are not tumour or tissue specific. A variety of other non-enteric neoplasms and non-malignant diseases of the colon, pancreas and hepatobiliary tree are associated with increases in circulating CEA. Evidence has accumulated indicating that in this group of patients the substances released into the circulation may either be CEA or a "CEA-like" substance(s). The problem of whether or not the material found in tissues and in sera of patients with non-enteric CEA cancers is identical to the CEA isolated from digestive system cancer or whether they are "CEA-like" substances that mimic the presence of CEA in sensitive radioimmunoassay systems remains to be fully elucidated. Furthermore, there are a substantial number of false negative results in colorectal tumours which limit the usefulness of the assay in primary tumour diagnosis and screening except in individuals at high risk to develop cancer.

It is therefore apparent, that the CEA radioimmunoassay in its current stage of development is not tumour-specific but merely detects, in a high incidence, a tumor-associated product(s). The separation of CEA and "CEA-like" molecules from a variety of tumour types and tissues, and a better understanding of the molecular structure of CEA may open the door to the establishment of more specific assays for malignancy.

CEA AS A TUMOUR MARKER IN COLORECTAL CANCERS

Although the CEA radioimmunoassay cannot be used in primary tumour diagnosis because of the relatively high frequencies of both false positive and false negative reactions, CEA measurement does play a role in cancer management, particularly in patients with established colorectal cancers and for this reason, its potential clinical usefulness in this context will be discussed in detail.

A number of studies have indicated that pre-operative CEA values are of some prognostic significance (17). Thus, if CEA levels are measured in the pre-operative state in established and previously diagnosed colorectal tumours, and if CEA is either undetectable or is present at levels less than 2.5 ng/ml, the prognosis appears to be better than when the levels are greater than 5 ng/ml. Furthermore, if the levels of CEA are measured in the pre-operative period and in the post-operative period, return of CEA levels in the post-operative state to the normal range is indicative of the complete resection of the primary tumour tissue (18). Failure of the elevated pre-operative CEA levels to return to the normal range after surgery would thus be an indicator of metastatic tumour spread even though at surgery it was thought that complete resection of the tumour tissues ostensibly occurred. In this context, the pre-operative and post-operative measurement of CEA levels is an excellent prognostic indicator both in the short and the longer term.

Perhaps the greatest role for CEA measurement in the clinical situation can be derived from its serial measurement during the course of disease in a patient with established colorectal carcinoma. Thus, the serial measurement of CEA in patients whose CEA level was elevated pre-operatively (80% of cases) can be used as an effective guide to the presence or absence of tumour recurrence. As noted previously, the decline of elevated CEA levels to the normal range following surgery shows a good correlation with a complete surgical resection of the tumour and potential surgical cure. Serial CEA measurements at approximately three month intervals during the first two years post-operatively and at more lengthy intervals thereafter is an effective guide to the presence of early tumour recurrence. Thus, the elevation of CEA in the absence of clinical manifestations frequently heralds the presence of tumour recurrence from three to thirty-six months in advance of clinical evidence of metastases (19,20). One must, however, be cautious about the interpretation of a single CEA value as transient CEA elevations can occur following resection of colorectal cancer. Thus, to document tumour recurrence by measuring CEA in a serial fashion, one requires the demonstration of at least two progressive elevations in CEA values at one month intervals. This presumptive evidence of tumour recurrence should then be confirmed, if possible, by conventional diagnostic imaging or endoscopic techniques. The early detection of tumour recurrence in this fashion could potentially result in the early initiation of chemotherapy and/or immunotherapy at a time when the total tumour burden is small and when such treatment is most likely to be effective. Unfortunately, current chemotherapy and immunotherapy for colorectal cancer is not particularly effective. However, looking into the future, when effective treatment for colorectal cancer will become available, the serial measurement of CEA should prove a most useful tool in the management of colorectal cancer patients. In addition, a negative CEA test, obtained serially in the post-operative state, is reassuring to the physician and patient constantly concerned with the possibility of tumour recurrence.

It should be recognized, however, that in about 20% of the patients, the CEA levels will not be elevated in the pre-operative period. In many of these patients, early tumour recurrence will again be indicated by a rise in CEA if serial studies are performed. In the minority of such patients, the CEA levels will remain normal throughout the patient's course even though extensive metastatic spread of tumour has occurred. In such patients, other biochemical parameters must be used to establish the early recurrence of disease.

SECOND LOOK SURGERY

More recently, serial elevations of CEA concentrations as a means of predicting early recurrence of colorectal cancer has been recommended as an indicator for "second look" abdominal surgery in that it might be an effective means of "curing" the patients (21). Even though CEA appears to be the most effective indicator of early tumour recurrence, the question then arises as to which patients will benefit from such a "second look" procedure. Clearly, in a small number of patients, an elevated CEA value will indicate the development of a second primary tumour and such individuals should, without question, be operated upon for cure. In the majority of instances, an elevated CEA value in the postoperative course of a patient in whom the CEA values were previously normal will reflect a recurrent growth of the original tumour. It is then important to determine, by conventional diagnostic imaging and endoscopic techniques, if this recurrence is localized or widespread in distribution as it is in only the former group that a "second look" surgical procedure is indicated. In some of these patients, it may not be possible to demonstrate the site of tumour recurrence by conventional techniques. Two potential approaches to resolve such difficulties can be considered. Computerized analysis of serial CEA levels in colorectal cancer patients has indicated that the rate of change in the circulating CEA concentration with time can provide useful information as to the nature of tumour regrowth. It appears that circulating CEA levels change at a maximal rate when hepatic metastases develop, presumably because the liver, which is largely responsible for CEA catabolism, no long functions effectively in this capacity. On the other hand, in local recurrence of the disease without hepatic involvement, a shallow rise in CEA concentration is observed with time (22).

Recently an exciting means of localing CEA-producing neoplasms has been developed through the use of isotopic imaging techniques employing radiolabelled purified anti-CEA antibody (23). Preliminary studies indicate that both primary and metastatic colorectal tumour lesions can be visualized by this novel, non-invasive method. This method would obviously be a most useful adjunctive diagnostic procedure to identify the extent and location of tissue in those patients considered for a "second look" procedure.

CEA AS A MONITOR OF THERAPY

In order for a tumour marker to be an effective monitor of therapy, it must accurately reflect the total tumour load at a given time. In many patients with malignancy, left untreated, the CEA levels are elevated but may change little even though the tumour continues to grow and spread. This would tend to indicate that the correlation of circulating CEA levels with tumour load is not an absolute one. However, a number of studies have indicated that CEA levels accurately reflect diminished tumour load following effective response to chemotherapy and radiation therapy (24,25). Failure of CEA levels to fall in response to treatment suggests that the tumour is resistant to that mode of therapy.

Ideally, multiple tumour markers should be used to monitor the response of tumour therapy. This has been clearly shown in teratocarcinoma, where simultaneous measurement of HCG and alpha-fetoprotein (AFP) levels more accurately reflect the response of these germ cell tumours to radiation and chemotherapy. Examples were cited where one or the other of these two markers would fall with therapy, but the other tumour product remained elevated, indicating a poor response to the mode of treatment that was initiated (26). It would therefore appear that on the basis of changes in the circulating levels of one tumour marker, one could erroneously conclude that a beneficial response of the tumour to treatment was observed. One could therefore envision that in the future, a battery of tumour markers will be successfully employed to monitor both the course and the therapeutic responses of a variety of tumours. Preliminary clinical studies with an isoenzyme of galactosyl transferase indicate the usefulness of this parameter in the management of colorectal cancer patients (27). Large scale clinical trials will require the development of a radioimmunoassay to replace the present assay of enzyme function which limits the number of samples that can be analyzed. However, it would appear that CEA and galactosyl transferase together may provide more accurate information with respect to diagnosis, recurrence and response of colorectal cancers to treatment.

An experimental, yet interesting, means of assessing the therapeutic responses of colorectal tumours involves the previously described technique of radiolabelled anti-CEA antibody. Following administration of isotopically administered anti-CEA immunoglobulin, the tumour can be localized and the persistence of the isotopic tracer used as a guide to the success or failure of treatment. More recently, large doses of I^{131}-rabbit anti-CEA immunoglobulin have been used to effectively treat a single patient with an intrahepatic cholangiocarcinoma. Although one cannot draw any conclusions from a single case study, the use of antibody to localize and deliver large doses of high energy radioisotopes to tumour tissue for therapeutic purposes is intriguing (28). The situation is analogous to the use of I^{131}, in doses of 50-125 millicuries, to treat thyroid cancer.

CEA IN OTHER TUMOURS

The foregoing discussion applies equally well to tumours other than those of the colon and rectum. Unlike colorectal tumours, the incidence of elevated CEA levels in localized disease is very low (29). However, with tumour progression, an increased incidence of CEA levels are observed. For example, in breast cancer, CEA levels are elevated in 15-30% of cases with localized disease and 70% of patients with metastatic involvement. Similar observations apply to lung, ovarian, genitourinary and other GI neoplasms. With respect to medullary carcinoma of the thyroid gland, thyrocalcitonin and CEA are two useful supplementary markers (29).

MULTIPLE TUMOUR MARKERS

To date, the data indicate that no single marker thus far described can be considered tumour specific. Clinically, these markers cannot be used in population screening and diagnosis, except perhaps in high risk groups within the population. Currently available markers are best employed serially to monitor the patient's course and response to treatment. Furthermore, no single marker adequately reflects tumour burden, and thus multiple tumour markers with varying degrees of tumour specificity, are required to follow and gauge the biology of host-tumour interaction.

REFERENCES

1. Foley, E.J. Cancer Res. 13:835, 1953.
2. Baldwin, R.W. In: The Physiopathology of Cancer. Ed. F. Homberger, Karger, Basel, 1:334, 1974.
3. Oettgen, H.F. and Hellstroem, K.E. In Cancer Medicine, Eds. J. F. Holland and E. Frei, pp. 951-990, Lea and Febiger, Philadelphia, 1973.
4. Herberman, R.B. Am. J. Clin. Path. 68:688, 1977.
5. Gold, P. and Freedman, S.O. J. Exp. Med. 121:429, 1965.
6. Gold, P. and Freedman, S.O. J. Exp. Med. 122:467, 1965.
7. Gold, P., Krupey, J., and Ansari, H. J. Natl. Cancer Inst. 45:219, 1970.
8. Rosenthal, K.L., Palmer, J.L., Harris, J.A., Rawls, W.E., and Tomkins, W.A.F. J. Immunol. 115:1049, 1975.
9. Fuks, A., Banko, C., Shuster, J., Freedman, S.O., and Gold, P. Biochem Biophys. Act 417:123, 1975.

10. Terry, W.D., Henkart, P.A., Coligan, J.E., and Todd, C.W. Transplant. Rev. 20:100, 1974.
11. Gold, P., Shuster, J., and Freedman, S.O. Cancer 42:1399, 1978.
12. Gold, P. CEA Group Report, International Research Group for Carcinoembryonic Proteins Meeting, Copenhagen, 1977 Scand. J. Immunol. (in press).
13. Krantz, M., Ariel, N., and Gold, P. In Proceedings of International Research Group for Carcinoembryonic Proteins Meeting, Marburg, 1978. Excepta Medica, (in press).
14. Hansen, H.J., Snyder, J.J., Miller, E., Vandevoorde, J.P., Neal Miller, O., Hines, L.R., and Burns, J.J. Human Pathol. 5:159, 1974.
15. Hirai, H. Cancer Res. 37:2267, 1977.
16. Lawrence, D.J.R., Stevens, V., Bettelheim, R., Darcy, D., Leese, C., Turberville, C., Alexander, P., Johns, E.W., and Neville, A.M. Br. J. Med. J iii:605, 1972.
17. Chu, T.N., Holyoke, E.D., and Murphy, G.P. New York State J. Med. 47:1388, 1974.
18. Shuster, J., Livingstone, A., Banjo, C., Silver, H.K.B., Freedman, S.O., and Gold, P. Am. J. Clin. Path. 62:243, 1974.
19. Mach, J.P., Vienny, H., Jaeger, P., Haldenmann, B., Egely, R., and Pettavel, J. Cancer 42:1439, 1978.
20. Sorokin, J.J., Sugarbaker, P.H., Zamchek, N., Pisick, M., Kupchik, H.Z., and Moore, F.D. JAMA 228:49, 1974.
21. Minton, J.P., and Martin, E.Q., Jr. Cancer 14:1422, 1978.
22. Staab, H.J., Anderer, F.A., Stumpf, E., and Fischer, R. Deut. Med. Woch. 102:1, 1977.
23. Goldenberg, D.M., Deland, F., Kim, E., Bennett, S., Primus, F.J., van Nagell, J.R., Estes, N., DeSimone, P., and Rayburn P. New Eng. J. Med. 298:1384, 1978.
24. Mayer, R.J., Gaorick, M.B., Steele, G.D., and Zamcheck, N. Cancer 42:1428, 1978.
25. Sugarbaker, P.H., Bloomer, W.D., Corbett, E.D., and Chaffey, J.T. Cancer 42:1434, 1978.
26. Lange, P.H., McIntire, K.R., Waldmann, T.A. New Eng. J. Med. 295:1237, 1976.
27. Podolsky, D.K., Weiser, M.M., Isselbacher, K.J., and Cohen, A.M. New Eng. J. Med. 299:703, 1978.
28. Ettinger, P.S., Dragon, L.H., Klein, J., Sgagias, M., and Order, S. Cancer Treatment Reports (in press)
29. Proceedings of the First International Conference on the Clinical Uses of Carcinoembryonic Antigen, Lexington, Kentucky, 1977, in supplement to Cancer 42:1397-1659, 1978.

THE STRUCTURE OF CARCINOEMBRYONIC ANTIGEN AND A GENETICALLY RELATED MATERIAL

J. E. Shively and C. W. Todd

Division of Immunology, City of Hope National Medical Center, Duarte, California 91010

Carcinoembryonic antigen (CEA) is one of the most widely studied tumor markers. It was first described by Gold and Freedman[1] as a component of adult colonic tumors and fetal gut. Subsequently it was identified in normal adult colonic mucosa[2,3] and in various noncolonic tumors[4,5,6]. Antigens which crossreact with CEA are NCA, found in normal lung and spleen,[7,8] BGPI, found in bile,[9] and TEX, found in liver metastases of colonic adenocarcinoma.[10] The presence of CEA or CEA crossreacting antigens in normal tissues helps to explain why circulating levels of CEA may be elevated in a number of inflammatory diseases and in various noncolonic cancers. This paper summarizes recent progress in structural studies on CEA and TEX, a genetically related material. An understanding of the structural relationships between CEA and related antigens should have direct application to the development of a more specific test for the diagnosis of various types of cancer and in monitoring cancer therapy.

IMMUNOLOGICAL STUDIES

CEA and TEX crossreact in double diffusion experiments. When tested against anti-CEA, CEA spurs over TEX; when tested against anti-TEX, TEX spurs over CEA.[10] In a competitive radioimmunoassay utilizing anti-TEX and radiolabeled TEX, about 300 ng of CEA is required to give 50% inhibition in the assay. In a similar assay utilizing anti-CEA and radiolabeled CEA, over 200 ng of TEX is required for 50% inhibition. Since these assays are capable of detecting as little as 0.5 ng of antigen, the amount of crossreactivity observed has very little effect on the specificity for each antigen.

A comparison of CEA isolated from tumor with CEA isolated from normal colon washings (NCW) on immunodiffusion shows complete identity whether tested with anti-CEA or anti-NCW.[2] TEX isolated from tumor and NCA isolated from normal spleens also show complete identity whether tested with anti-NCA (Figure 1A) or anti-TEX (Figure 1B). However, anti-TEX reacts differently with CEA than does anti-NCA. No precipitin line is formed between anti-NCA and CEA (Figure 1A), but one is formed between anti-TEX and CEA (Figure 1B). In a similar fashion, anti-CEA raised in monkeys, a highly specific antisera for CEA,[11] will crossreact with TEX but not with NCA on immunodiffusion (Figure 1C). These findings suggest that TEX possesses an antigenic determinant not found in NCA but found in CEA. A radioimmunoassay with high specificity for NCA has been developed in this laboratory[7] and in others.[12,13]

AMINO ACID AND CARBOHYDRATE COMPOSITIONS

Table I presents the amino acid and carbohydrate compositions of CEA, NCW, TEX, and NCA. The degree of relatedness is obvious. Neither CEA nor NCW contains methionine. TEX and NCA contain about 4-5 moles of methionine per mole of glycoprotein. TEX and NCA also differ from CEA and NCW in terms of total carbohydrate content. They have roughly half as much carbohydrate. The change in carbohydrate composition appears to be mainly due to a decreased amount of N-acetylglucosamine in TEX and NCA compared to CEA and NCW. The molecular weight differences can be accounted for by the changes in per cent carbohydrate. Thus, the calculated molecular weights for the polypeptide chains of all four materials is about 70,000 daltons.

Table II shows the striking homology in the amino terminal sequences of CEA, NCW, TEX, and NCA. They differ only at position 21 where valine is found in CEA or NCW, but alanine in TEX or NCA. Clearly, all four glycoproteins are related, but CEA and NCW can be grouped together and TEX and NCA can be grouped together as most closely related, if not identical pairs.

It is interesting to speculate on the relationship between CEA and its crossreacting antigens. An ancestral gene probably has undergone gene replication and subsequent mutations. The resulting gene products would have closely similar polypeptide chains, however there would be various amino acid substitutions, such as that observed at residue 21 in the amino terminal sequences. This evolutionary divergence also explains the occurrence of methionine in TEX or NCA but not in CEA or NCW. The new gene products might well be differentially expressed in various tissue, hence the finding of CEA or NCW in the colon and NCA in the spleen or lung. Perhaps their expression would also vary during differentiation. Thus, CEA is produced in larger amounts in fetal or tumor tissue, but NCA may be produced more uniformly in certain adult tissues. Furthermore, the gene products may be glycosylated differently. This could be due to the altered amino acid sequence or perhaps a tissue specific difference in glycosylation sequences. A major unanswered question centers on whether or not the glycosylation difference is important to the function of these molecules. Since their functions are still unknown, one can only speculate about the role of glycosylation.

CEA STRUCTURE

AMINO ACID SEQUENCE OF TRYPTIC FRAGMENTS FROM CEA

Due to its high degree of glycosylation CEA is very resistant to proteolytic cleavage. However, in the presence of Triton X100 the extent and uniformity of cleavage improved considerably. The resulting peptides were purified by a combination of ion exchange-high pressure liquid chromatography and gel filtration. The amino terminal sequences obtained are shown in Table III. In all cases the amino terminal sequences came to an abrupt halt during the course of Edman degradation. It is likely that carbohydrate substitution prevented extended sequencing. In the case of peptide number 7, the sequence Asn X Thr is a known recognition sequence for glycosylation. Thus, it is possible that at least in a portion of CEA molecules this Asn residue is glycosylated. These studies convined us that before further structural information could be obtained on CEA, it would be necessary to first deglycosylate CEA. Presumably CEA free of carbohydrate would be more amenable to the usual techniques employed in protein structural studies.

DEGLYCOSYLATION OF CEA

Two chemical methods for the deglycosylation of CEA were investigated. One utilized solvolysis in anhydrous HF as described by Mort and Lamport,[17] and the other treatment with methanolic HCl under mild conditions. HF treatment of 82 mg of CEA yielded 48 mg of deglycosylated material after purification by gel filtration.[18] Since HF treatment has been shown to remove all O-glycosidically linked carbohydrate but not N-glycosidically linked carbohydrate, the deglycosylated product should contain one residue of N-acetylglucosamine per asparagine which was originally glycosylated:

```
        -NHCHCO-
           |
          CH2                 CH2OH
           |                    |
          CONH ──── HO ─┐     ┌─
                        \___ /
                        /    \
                       /      OH
                      NHCOCH3
```

In the case of CEA, the deglycosylated product should contain about 17% N-acetylglucosamine, and the expected yield of product would be 49 mg. The amino acid and carbohydrate analyses for HF deglycosylated CEA are shown in Table IV. These results show that HF treatment has little or no effect on the amino acid composition of CEA and effectively removes all of the carbohydrate except for the N-glycosidically linked N-acetylglucosamine.

The HF deglycosylated product when subjected to automatic Edman degradation gave a single NH_2-terminal sequence in high absolute yield (71%) and high repetitive yield (98%). This demonstrates that HF treatment had no effect on the NH_2-terminus of CEA, did not cause peptide bond cleavage resulting in the formation of new NH_2-termini, and produced a protein more amenable to extended sequencing than intact CEA.

HF deglycosylated CEA is only sparingly soluble in water, but retains 15% by weight of its original antigenic activity as measured by radioimmunoassay and inhibits up to 85% of the binding of intact CEA to anti-CEA. The solubility of HF treated CEA can be increased by performic acid oxidation, thus suggesting that HF induces disulfide bond interchange. Performic acid oxidized CEA retains little or no antigenic activity when tested against anti-intact CEA.

CEA deglycosylated by mild methanolysis yields a material with even less carbohydrate than HF treated CEA. The analyses shown in Table IV suggest that methanolysis has no effect on the amino acid composition of CEA. However, SDS polyacrylamide gel electrophoresis reveals that considerable peptide bond cleavage occurs. Methanolyzed CEA is soluble in water but has little or not residual antigen activity when tested against anti-intact CEA.

CEA deglycosylated by mild methanolysis yields a material with even less carbohydrate than HF treated CEA. The analyses shown in Table IV suggest that methanolysis has no effect on the amino acid composition of CEA. However, SDS polyacrylamide gel electrophoresis reveals that considerable peptide bond cleavage occurs. Methanolyzed CEA is soluble in water but has little or no residual antigenic activity.

More recent studies have focused on the production, separation, and sequencing of specific peptides from deglycosylated CEA. These techniques will also be applied to the study of CEA-related antigen, TEX.

ACKNOWLEDGEMENTS

This research was supported by the National Cancer Institute through grant 16434 and by grant 19163 from the National Large Bowel Cancer Project.

REFERENCES

1. Gold, P., Freedman, S.O.: Demonstration of tumor-specific antigens in human colonic carcinomata by immunological tolerance and absorption techniques. J. Exp. Med. 121:439-462, 1965.

2. Egan, M.L., Pritchard, D.G., Todd, C.W., Go, V.L.W.: Isolation and immunochemical and chemical characterization of carcinoembryonic antigen-like substances in colon lavages in healthy individuals. Cancer Res. 37:,2638-2643, 1977.
3. Martin, F., Martin, M.S.: Demonstration of antigens related to colonic cancer in the human digestive system. Int. J. Cancer 6:352-360, 1970.
4. DeYoung, N.J., Ashman, L.K.: Physicochemical and immunochemical properties of carcinoembryonic antigen (CEA) from different tumor sources. Aust. J. Exp. Biol. Med. Sci. 56:321-331, 1978.
5. Ishikawa, N., Hamada, S.: Association of medullary carcinoma of the thyroid with carcinoembryonic antigen. Br. J. Cancer. 34, 111-115, 1976.
6. Denk, H., Tappeiner, G., Eckerstorfer, R., Holzner, J.H.: Carcinoembryonic antigen (CEA in gastrointestinal and extra-gastrointestinal tumors and its relationship to tumor-cell differentiation. Int. J. Cancer 10:262-272, 1972.
7. Engvall, E., Shively, J., Wrann, M,: Isolation and characterization of the normal crossreacting antigen: Homology of its NH_2-terminal amino acid sequence with that of carcinoembryonic antigen. Proc. Nat. Acad. Sci. (USA) 75:1670-1674, 1978.
8. von Kleist, S., Chavanel, G., Burtin, P.: Identification of an antigen from normal human tissue that crossreacts with the carcinoembryonic antigen. Proc. Nat. Acad. Sci. (USA) 69: 2492-2494, 1972.
9. Svenberg, T.: Carcinoembryonic antigen-like substances of human bile. Isolation and partial characterization. Int. J. Cancer, 17:588-596, 1976.
10. Kessler, M.J., Shively, J.E., Pritchard, D.G., Todd, C.W.: Isolation, immunological characterization, and structural studies of a tumor antigen related to carcinoembryonic antigen. Cancer Res. 38:1041-1048, 1978.
11. Ruoslahti, E., Engvall, E., Vuento, M., Wigzell, H.: Monkey antisera with increased specificity to carcinoembryonic antigen (CEA). Int. J. Cancer 17:358-361, 1976.
12. Newman, E.S., Petros, S.E., Georgiadis, A., Hansen, H.J.: Interrelationship of carcinoembryonic antigen and colon carcinoma antigen-III. Cancer Res.34:2125-2130, 1974.
13. von Kleist, S., Troupel, S., King, M., Burtin, P.: Étude comparative du taux sérique du NCA et du CEA. Bull. du Cancer 63:627-632, 1976.
14. Terry, W.D., Henkart, P.A., Coligan, J.E., Todd, C.W.: Structural studies of the major glycoprotein in a preparation with carcinoembryonic antigen activity. J. Exp. Med. 136:200-204. 1972.
15. Shively, J.E., Todd, C.W., Go, V.L.W., Egan, M.L.: Amino-terminal sequence of a carcinoembryonic antigen-like glyco-

protein isolated from the colonic lavages of healthy individuals. Cancer Res. 38:503-505, 1978.
16. Shively, J.E., Kessler, M.J., Todd, C.W.: The amino-terminal sequences of the major tryptic peptides obtained from carcinoembryonic antigen by digestion with trypsin in the presence of Triton X100. Cancer Res. 38:2199-2208, 1978.
17. Mort, A.J., Lamport, D.T.A.: Anhydrous hydrogen fluoride deglycosylates glycoproteins. Anal. Biochem. 82:289-309, 1977.
18. Glassman, J.N.S., Todd, C.W., Shively, J.E.: Chemical deglycosylation of carcinoembryonic antigen for amino acid sequence studies. Biochem. Biophys. Res. Commun. 85:209-216, 1978.

TABLE I. AMINO ACID AND CARBOHYDRATE COMPOSITIONS

	Mole Percent			
	CEA	NCW	TEX	NCA
Tyrosine	4.0	5.4	4.0	4.3
Phenylalanine	2.2	3.0	2.9	1.7
Lysine	2.9	3.2	3.2	3.2
Histidine	1.7	1.9	1.4	1.2
Arginine	3.1	3.1	2.7	2.9
Aspartic acid[1]	15.7	13.4	13.8	13.7
Threonine	8.7	8.0	9.2	8.5
Serine	11.4	10.2	9.9	10.0
Glutamic acid[1]	9.6	9.6	10.9	11.6
Proline	8.0	9.4	8.2	8.1
Glycine	6.3	5.2	6.4	6.6
Alanine	6.1	5.5	6.4	5.9
Cysteine	1.6	1.7	1.3	0.8
Valine	6.6	6.3	5.9	6.9
Methionine	0	0	0.9	0.9
Isoleucine	3.7	4.3	4.0	4.0
Leucine	7.6	8.4	9.2	8.3
	Weight Percent			
Fucose	8.1	7.2	6.0	3.1
Mannose	6.8	7.5	9.4	10.5
Galactose	10.5	7.9	6.5	3.6
N-Acetylglucosamine	20.5	20.9	8.9	11.0
Neuraminic acid	3.2	-	1.7	2.9
Amino acids[2]	40	50	65	70
Carbohydrate[2]	60	50	35	30
Molecular Weight[3]	180K	170K	110K	100K

[1] Represents sum of acid and amide.
[2] Percent of total recovered.
[3] Based on gel filtration.

TABLE II. NH$_2$- Terminal Sequences

Position	CEA[1]	NCW[2]	TEX[3]	NCA[4]
1	Lys	Lys	Lys	Lys
2	Leu	Leu	Leu	Leu
3	Thr	Thr	Thr	Thr
4	Ile	Ile	Ile	Ile
5	Glu	Glu	Glu	Glu
6	Ser	Ser	Ser	Ser
7	Thr	Thr	Thr	Thr
8	Pro	Pro	Pro	Pro
9	Phe	Phe	Phe	Phe
10	Asn	Asn	Asn	Asn
11	Val	Val	Val	Val
12	Ala	Ala	Ala	Ala
13	Glu	Glu	Glu	Glu
14	Gly	Gly	Gly	Gly
15	Lys	Lys	Lys	Lys
16	Glu	Glu	Glu	Glu
17	Val	Val	Val	Val
18	Leu	Leu	Leu	Leu
19	Leu	Leu	Leu	Leu
20	Leu	Leu	Leu	Leu
21	Val	Val	Ala	Ala
22	His	(His)	His	His
23	Asn	Asn	Asn	Asn
24	Leu	Leu	Leu	Leu

[1] Terry *et al* 14
[2] Shively *et al* 15
[3] Kessler *et al* 10
[4] Engvall *et al* 7

TABLE III. NH$_2$-Terminal Sequences of CEA Tryptic Peptides[1]

Peptide	1	2	3	4	5	6	7	8	9	10	11	12	13	14	15
1	ser	leu	ser	ile	phe	cys	val	asn	tyr	asn	gln	phe	val	gln	tyr
2	ile	asp	ile	ile	phe	val	val	asp	ala	ala	thr	phe	ala	gln	gly
3	leu	ile	pro	leu	tyr	gly	–	–	thr	tyr					
4	ile	leu	val	gln	pro	leu	val	thr	gln	ala	asp	glu	–	ala	phe
5	ala	gly	ile	val	phe	ala	ser	ile	tyr	glu	gly	gly			
6	lys	ile	thr	val	gly	ala	asn	gly	asp						
7	leu	asp	ser	thr	val	gly	asn	glu	thr	ala	ser	gly	glu		
8	ser	leu	val	val	gln	ala	glu	leu	val	lys	–	–	–	leu?	
9	leu	glu	val	asn	ser	ala	glu	ser	val	leu					
10	asp	asp	gly?	lys	pro	–	val	gly							

[1]Taken from Shively *et al*[16]

TABLE IV. AMINO ACID AND CARBOHYDRATE COMPOSITION OF DEGLYCO-
SYLATED CEA

	Mole Percent		
		After	
	Before	HF Treatment	Methanolysis
Tyrosine	3.5	4.7	4.5
Phenylalanine	1.5	2.3	2.5
Lysine	2.8	2.5	2.5
Histidine	1.7	1.6	1.7
Arginine	3.7	4.2	3.8
Aspartic Acid	16.0	14.5	15.1
Threonine	9.4	9.3	9.8
Serine	12.0	11.4	11.0
Glutamic Acid	10.9	10.8	10.9
Proline	11.0	10.7	9.8
Glycine	6.7	5.9	5.6
Alanine	6.4	6.3	6.8
Valine	4.5	5.3	5.4
Methionine	0.0	0.0	0.0
Isoleucine	3.8	3.0	2.8
Leucine	6.4	7.4	7.1
	Weight Percent		
Glucosamine	16.3	6.6	3.9
Mannose	9.0	1.0	0.4
Galactose	9.4	0.0	0.0
Gucose	8.5	0.0	0.0
Neuraminic Acid	1.8	0.0	0.0

Figure 1. Comparison of CEA, TEX, and NCA on immunodiffusion

A: 1 and 4, TEX; 2, NCA; 3 and 5, CEA; 6, goat anti-NCA

B: 1 to 5 are the same as in A; 6, rabbit anti-TEX.

C: 1 to 5 are the same as in A; 6, monkey anti-CEA.

ONCODEVELOPMENTAL ISOENZYMES

William H. Fishman

La Jolla Cancer Research Foundation, 2945 Science Park Road, La Jolla, California 92038

It is important to establish the legitimate level of expectation for the clinical value of tumor markers. Four examples of the most satisfactory tumor markers to date will illustrate several of the fundamental considerations.

HCG production by trophoblastic tumors is a characteristic which provides the physician with a specific diagnostic aid and an objective measure of tumor response. The outstanding example of this is the story of methotrexate treatment of choriocarcinoma introduced by Li, et al.[1] Trophoblast cells normally produce excessive amounts of HCG, and this is a phenotype of the malignant trophoblast cell.

Other tumors with trophoblastic elements such as testicular teratocarcinoma also produce HCG and the blood levels are of real clinical utility to physicians.

Another example of the exaggerated synthesis in tumors of a protein characteristic of the cell of origin is prostatic acid phosphatase.[2] Although an elevated prostatic acid phosphatase in the serum is considered to be pathognomonic for cancer of the prostate, the test is still far from infallible. The advantage is that a positive test in a male alerts the physician to examine a specific organ site. The disadvantage is that the test is usually positive after the tumor has spread through the capsule. Too many cancers of the prostate don't produce enough acid phosphatase to cause a serum rise. There does, however, appear to be an improvement in clinical yield by the introduction of an RIA procedure.[3]

The very best tumor marker I know is calcitonin in medullary thyroid cancer.[4] The calcitonin producing cells of the thyroid are the so-called C-cells which in embryonic life originated from the neural crest and migrated to the thyroid. In medullary thyroid cancer, urinary calcitonin is invariably elevated. It has been possible to identify pre-clinical medullary thyroid cancer by provoking calcitonin excretion. However, the calcitonin pre-cancer test is effective only in families which are known to carry a predisposition for medullary thyroid cancer. Removal of the thyroid glands is an effective prevention measure in these individuals for contracting medullary thyroid cancer.

Supported by grants in aid CA-22384 and R01-21967 of the National Cancer Institute, National Institutes of Health, Bethesda, Md.

Finally, the correlation of AFP production with yolk sac tumor of the testis[5] is excellent and with hepatoma[6] good. Yet these are rare tumors.

From this review of the most specific tumor markers, the following statements can be made.
  i. All are valuable in defined <u>different</u> populations at high or above average risk for a specific type of cancer.
  ii. All are valuable in the differential diagnosis of cancer.
  iii. None are cost effective for screening whole populations.

From my point of view, these tumor markers fit the definition of cancer as a disease of gene regulation and provide logic to the exploration of this field as a source of other tumor markers.

ONCODEVELOPMENTAL ALKALINE PHOSPHATASES[7,8]

In this presentation, my main subject will be a family of developmental alkaline phosphatases which illustrates the advantages to the investigator of a protein with a catalytic site.

As you know, a clinical problem of real importance in the 1950's was to distinguish liver and bone alkaline phosphatases from each other in the serum of patients. In approaching this problem, we developed a scheme of fractionating alkaline phosphatase isoenzymes (Table I) which recognized four varieties of the enzyme: liver, bone, intestine and placenta.

TABLE I.   ISOENZYMES OF AKALINE PHOSPHATASE

Properties	Plac.	Int.	Liv.	Bone
L-Phenylalanine-inhibition(%)	75	75	0-10	0-10
L-Homoarginine-inhibition(%)	5	5	78	78
Heat inactivation(%)	0	50-60	50-70	90-100
Order of electrophoretic migration (anodal)	3	4	1	2
Hydrolysis of neuraminidase	+	0	+	+
Cross-reactivity				
with placental isoenzyme antisera	+	0	0	0
with intestinal isoenzyme antisera	0	+	0	0
with liver isoenzyme antisera	0	0	+	+

REGAN ISOENZYME[9]

In 1968, my group at Tufts University School of Medicine discovered the presence of term placental alkaline phosphatase isoenzyme in the serum of a man with disseminated bronchogenic cancer. When the tumor was studied later, cytochemical evidence was obtained for the placental isoenzyme in the cytoplasm of the cancer cells. This isoenzyme was named the Regan isoenzyme. This observation arrested our attention. Why were cancer cells in a man making a placental protein? We are still trying to find the answer to this question.

There is diagnostic potential, however, in the appearance of an ectopic protein in the blood of a cancer patient. The clinical studies which have been done since[10,11,12] demonstrate that 15 to 25% of all cancer patients exhibit Regan isoenzyme. (There is surprisingly a number of benign proliferative disorders which also express Regan isoenzyme usually at low levels.) In those cancer patients with elevated levels, successful therapy is accompanied by a drop in serum level and recurrence is often preceded by a rise in serum Regan isoenzyme.

In Japan, two years later, a Regan variant was reported as the Nagao isoenzyme.[13] Like Regan, it was heat-stable and inhibited by L-phenylalanine but unlike Regan, it was specifically inhibited by L-leucine. This isozyme shares the L-leucine inhibition with a rare phenotype of placental alkaline phosphatase.

In subsequent studies we were able to show that the Regan isoenzyme was expressed as the three common F, FS and S electrophoretic phenotypes of normal placenta. It is of interest that the gene for placental alkaline phosphatase has 18 alleles which express 44 electrophoretic phenotypes.

NON-REGAN ISOENZYME

Tumors also produce alkaline phosphatase which is not term placental but which shares biochemical and immunologic properties with liver and bone isoenzymes, often in association with Regan isoenzyme.

In 1976, L. Fishman, et al.[14] discovered that early placental expressed the non-Regan phenotype and that by the end of the first trimester the Regan isoenzyme predominated.

Of great help in our experimental studies have been the model systems from two sublines of HeLa cells. One such line TCRC-1 produces only term placental alkaline phosphatase while another TCRC-2 produces the early placental isoenzyme. The properties of these two cell lines appear in Table II.

TABLE II. CONTRASTING BIOLOGICAL AND BIOCHEMICAL PROPERTIES OF TWO SUBLINES OF HELA CELLS.

	TCRC-1	TCRC-2
Produces Regan Isoenzyme	+	-
Produces non-Regan Isoenzyme	-	+
Alkaline Phosphatase		
Kinetic Properties		
Specific activity (μmoles/min/μg)	0.75	3.2
L-Phenylalanine inhibition (%)	73.1	0.0
L-Homoarginine inhibition (%)	11.5	77.5
Heat inactivation (%, 5 min. at 65C)	10.9	100.0
Immunological characteristics		
Placental determinants	+	-
Intestinal determinants	-	-
Liver determinants	-	+
Induction		
Prednisolone effect on specific activity (%)	+175	0
Acidic isoferritins		
Induction by inorganic iron	+	-
β-Glucuronidase	+++	+
Induction b- prednisolone	+	-
Growth in immunosuppressed rate	+++	-

Source: Can Res. 36:4256, 1976, used with permission.

KASAHARA ISOENZYME

The Kasahara isoenzyme possesses some of the properties of placental isoenzyme and was first observed in patients with hepatocellular carcinoma.[15] Higashino, et al. made an extensive study of this isoenzyme.[16]

OTHER ONCOTROPHOBLAST PROTEINS

We reasoned back in 1974 that if one placental protein is produced by cancer cells that other can be expected also. Previously human chorionic gonadotrophin (HCG), an entopic product, was associated with trophoblastic cancer, choriocarcinoma. We found that neoplastic ascites from ovarian cancer which contained Regan isoenzyme frequently expressed HCG. Other workers have found three placental proteins (Regan isoenzyme, placental alkaline phosphatase, and somatomammotropin) in certain malignancies.

In a number of longitudinal studies on patients with ovarian carcinoma, both concordance and non-concordance of expression of Regan isoenzyme and HCG were observed. This suggests a selection process of particular genotypes in the ovarian malignancies.

MODULATION OF EXPRESSION

From the early observations of Cox, McLeod, and Griffen that corticosteroid hormone in tissue culture medium enhanced the alkaline phosphatase activity has developed a tool for the study of gene expression. Its modulation is defined as a change in level of gene product.

As shown in Table II, prednisolone induces term alkaline phosphatase in the subline TCRC-1 which is monophenotypic for the term isoenzyme but not in TCRC-2 which is monophenotypic for early placental isoenzyme.

n-Butyrate is another modulator which often behaves like prednisolone but not in the Dot cervical cancer cell line.[1]

CORRELATION OF NEOPLASTIC AND DEVELOPMENTAL ISOENZYMES

Year	Neoplastic	Developmental
1968	Regan Isoenzyme	Term placental, F, FS, S phenotypes
1969	Non-Regan Isoenzyme	Early-placental phenotype
1970	Kasahara Isoenzyme	F1-amnion
1971	Nagao Isoenzyme	Term placental D-variant

GAMMA GLUTAMYL TRANSFERASE-GGT

When Dr. Krishnaswamy visited the Foundation in 1977, he brought to our attention the fact that seminal plasma was very rich in GGT. This suggested to us that perhaps seminoma patients' sera might be rich in GGT. Dr. McIntire collaborated with us by sending 25 sera from patients with seminoma. Nine of these exhibited elevated GGT. On reviewing the other data on these patients, it turned out that these nine produced HCG or AFP or both. These products could only be coming from chorionic or yolk sac elements of mixed germ cell tumors. These correlations suggested that GGT was possibly an oncodevelopmental enzyme. Other evidence which we have gathered demonstrates GGT in ovarian tissue, in early placenta, in teratocarcinoma and in ovarian carcinoma.

The gene products which appear in teratocarcinoma include early and term placental alkaline phosphatase, yolk sac derived AFP, GGT, chorionic gonadotropin, CEA and isoferritins.

DEVELOPMENTAL PERSPECTIVE

Simply collecting examples of oncodevelopmental proteins as such is apt to be less meaningful than attempting to relate them in the context of normal development.

For example, the HCG which is produced in development is measurable first at seven days of gestation, a stage corresponding to the implantation of the blastocyst. Although the major synthesis of HCG occurs during the first trimester, it continues throughout pregnancy.

In the trophoblast chronology, early placental alkaline phosphatase appears from 6 weeks and is superimposed by term placental alkaline phosphatase after 10 weeks.

AFP, on the other hand, first appears in the yolk sac which is an event occurring between 2 and 3 weeks after fertilization of the egg. AFP later becomes a phenotype of fetal liver which is the major producer of this oncodevelopmental protein during fetal life.

These five examples suggest that it may be possible to use the cDNA's which can be prepared from the mRNA's of specific gene products as a means of marking DNA sequences for particular oncodevelopmental genes.

SUMMARY

We view cancer as a disease of gene regulation and interpret the expression of developmental proteins as a reflection of the disorder. It is possible that ectopic synthesis of developmental proteins may contribute to the malignant phenotype. This possibility merits a great deal of investigation.

It may also be significant that the gene products of germ cell tumors are expressed in non-germ cell tumors, for example, HCG in teratocarcinoma and HCG in bronchogenic cancer.

We do predict that success in tumor marker studies should follow advanced in evaluating disorders in gene regulation as manifested by the ectopic expression of developmental proteins.

REFERENCES

1. Li, M.C., Hertz, R., Spencer, D.B.: Effect of methotrexate therapy upon choriocarcinoma and chorioadenoma. Proc. Soc. Exp. Biol. & Med. 93:361-366, 1956.
2. Fishman, W.H., Lerner, F.: A method for estimating serum acid phosphatase of prostatic origin. J. Biol. Chem. 200: 89, 1953.
3. Cooper, J.F., Foti, A.: A radioimmunoassay for prostatic acid phosphatase 1. Methodology and range of normal male serum values. Can. Res. 35:2446, 1975.
4. Melvin, D.E.W., Miller, H.H., Tashjian, A.H., Jr.: Early diagnosis of medular-carcinoma of the thyroid gland by means of a calcitonin assay. NEJM 285:1115-1120, 1971.
5. Kohn, J., Orr, A.H., McElwain, T.J., Bental, M., Peckham, M.J.: Serum alpha-fetoprotein in patients with testicular tumors. Lancet 2:433-436, 1976.
6. Abelev, G.I., Perova, S.D., Khramkova, N.I., Postnikova, Z.A., Irlin, I.S.: Embryonic serum alpha-globulin and its synthesis by transplantable mouse hepatomas. Biokhimiya 28: 625-634, 1963.
7. Fishman, W.H.: Perspectives on alkaline phosphatase isoenzumes. Amer. J. Med. 56:617, 1974.
8. Kottel, R.H., Fishman, W.H.: Developmental Alkaline Phosphatases as Biochemical Tumor Markers. Marcel, Dekker,Inc. (in press)
9. Fishman, W.H., Inglis, N.R., Stolbach, L.L., Krant, M.J.: A serum alkaline phosphatase isoenzyme of human neoplastic cell origin. Can. Res. 28:150, 1968.
10. Stolbach, L.L., Krant, M.J., Fishman, W.H.: Ectopic production of an alkaline phosphatase isoenzyme in patients with cancer. NEJM 281:457, 1969.
11. Nathanson, L., Fishman, W.H.: New observations on the Reggan isoenzyme of alkaline phosphatase in cancer patients. Cancer 27:1388, 1971.
12. Cadeau, B.J., Blackstein, M.E., Malkin, A.: Increased incidence of placenta-like alkaline phosphatase activity in breast and genitourinary cancer. Can. Res. 34:729, 1974.
13. Nakayama, T., Yoshida, M., Kitamura, M.: L-leucine sensitive heat-stable alkaline phosphatase isoenzyme detected in a patient with pleuritis carcinomatosa. Clin. Chim. Acta. 30:546, 1970.
14. Fishman, L., Inglis, N.R., Fishman, W.H.: Preparation of two antigens of human liver isoenzymes of alkaline phosphatase. Clin. Chim. Acta. 34:393, 1971.
15. Warnock, M.L., Reisman, R.: Variant alkaline phosphatase in human hepatocellular cancer. Clin. Chim. Acta 24:5, 1969.
16. Higashino, K., Kudo, S., Otani, R., Yamamura, Y., Honda, T. Sakurai, J.: A hepatoma-associated alkaline phosphatase, the

Kasahara isozyme, compared with one of the isozymes of FL amnion cells. Annal. New York Acad. Sci. 259:337, 1975.

ENZYMES IN HUMAN CANCER: FROM THE SPECIFIC TO THE MORE GENERAL

David M. Goldberg, M.D., Ph.D.

Professor and Chairman, Department of Clinical Biochemistry, The University of Toronto; and Biochemist-in-Chief, The Hospital for Sick Children, Toronto, Canada

INTRODUCTION AND SCOPE

The previous two speakers have dealt with highly selective aspects of enzymes in cancer and in so doing have brought us to the limits of our present knowledge in this area with appropriate emphasis on the novel discoveries of the last few years. The original theme I was requested to tackle by the organizers was, "Other Enzymes in Cancer." If I have consented to bear this burden of Atlas, it is only because I have subconsciously substituted the qualifying term "some" for "all" in defiance of a remit which was probably implied but never overtly stated in the invitation.

To define the ground rules of this presentation more succintly, I want to emphasize that diagnostic considerations will be paramount. This means that I shall be reviewing where we stand with tests that have been in the repertoire for several decades although newer interpretative and technical aspects will be emphasized where appropriate. Some fundamental studies involving these same enzymes will also be mentioned where they point to a possible function in the cancer cell. To begin with, however, I wish to outline some studies we have carried out on the enzymology of cancer of the human uterine cervix because these have uncovered a range of alterations which expand the mechanisms and concepts outlined by the previous two speakers.

ENZYMES IN CANCER OF THE HUMAN UTERINE CERVIX

1. <u>Nucleases</u>: More than a decade ago, we demonstrated dramatic increases in the ribo- and deoxyribo-nuclease content of biopsy samples of human uterine cervix cancer compared with normal cervix tissue (1). These observations are capable of several interpretations, one of which would emphasize the change in cell population in the nonmalignant component of the cancer biopsy sample due to reactive infiltration of the *lamina propria* with inflammatory cells. On the not yet proven assumption that the increased nucleases genuinely represent a feature of the malignant epithelium, it is reasonable to suggest that actively dividing cells turning over nucleic acids at a fast rate will require a suitable complement of catabolic as well as anabolic enzymes. Heterogeneity of these enzymes has also to be taken into account as well as their locus within the cell. The magnitude of the increases, especially for ribonuclease, and their expression in

more than one intracellular compartment (Table 1) lends some credence to the idea that the described changes may be important for the transition to neoplasia in this particular tissue.

In subsequent studies it was shown that radiation therapy altered the nuclease content of the cervical cancers in a disparate manner, deoxyribonuclease declining in activity and ribonuclease increasing (2). The higher ribonuclease activity could be detected in exfoliated cells isolated from vaginal washings (Figure 1), but there was no obvious correlation between the extent of this increase and the response to radiation therapy (3). At a more clinical level, it was demonstrated that increased ribonuclease activity occurred in the cellular material obtained from vaginal washings of patients with cervical cancer than from patients with nonmalignant gynecologic lesions (4). This was more reliable than cytology in diagnosing invasive cancer of the cervix but was of no value in detecting *in situ* lesions (5).

2. <u>Glycolytic Enzymes</u>: One of the earliest observations concerning the metabolic behaviour of cancers drew attention to their high rate of aerobic glycolysis compared with normal tissues (6). Accordingly, we directed a number of investigations into the content and characteristics of glycolytic enzymes in uterine cancer. A range of tissues were employed as normal controls and included fresh surgical biopsies as well as epithelial and sub-epithelial layers of *post mortem* cervix samples (7). Virtually all enzymes examined had significantly higher activities in the cancer tissue compared with the normal control material (Table 2). This increased activity occurred with regulatory as well as with non-regulatory enzymes of the glycolytic pathway and provides an obvious explanation for the increased glycolytic capacity in this particular cancer. In addition to changes in total activity, alterations in molecular composition were detected by isoenzyme analysis. One of the more notable changes was in hexokinase, where hexokinase 2, never detected by us in normal tissue of the uterine cervix, was present in over 30% of the samples from cancer patients (8). This is a more absolute difference between normal and malignant cervix in respect to hexokinase isoenzymes than has previously been reported (9).

Detailed studies on phosphoglucomutase isoenzymes were performed to classify the patient material according to phenotype at the PGM_1 gene locus (10). No significant difference in distribution of the phenotypes at this locus compared with that previously defined for the British population (11) was apparent for normal or malignant cervix samples, which in turn did not differ significantly from each other; by contrast, a highly significant preponderance of the PGM_1-1 phenotype with respect to normal endometrium and to the previously cited population statistics was observed in samples of endometrial cancer. Whether this implies that suppression of the PGM_1-2 gene product occurs in endometrial cancer, or that females who are phenotypically PGM_1-1 individuals are highly susceptible to endometrial cancer remains to be established, and is clearly a worthwhile objective. It is unfortunate that in the patients we have so far examined we did not take the opportunity to characterize the erythrocyte PGM isoenzymes to ascertain whether these matched the phenotype of uterine tissues.

Yet another aberration associated with cervical cancer was seen for the enzyme pyruvate kinase (12). As shown in Figure 2, the enzyme of normal cervix displays conventional hyperbolic substrate-activity relationships with only modest increase in activity in the presence of optimal concentrations of fructose diphosphate (FDP). By contrast, pyruvate kinase of cervical cancer shows sigmoidal substrate-activity relationships converted to the conventional hyperbolic pattern in the presence of FDP which therefore causes greatly enhanced activity at sub-saturating substrate concentrations (Figure 3).

As in our earlier study with nucleases, we analyzed the effects of radiation therapy upon glycolytic enzymes in cancers of the cervix and noted a sharp decline in the activity of most (13). Phosphofructokinase was the most profoundly affected (Figure 4) to the extent that the activity in post-radiation samples not only fell from the values encountered in pre-radiation samples, but were actually below those encountered in normal cervix samples. Since this enzyme is widely regarded as the rate-limiting enzyme of the glycolytic pathway, this dramatic reduction in its activity may seriously compromise carbohydrate metabolism of the cancer cell and thus account in part for the lethal effects of ionizing radiation on this tumour.

In summary, then, we have shown that studies of glycolytic enzymes in a specific type of tumour can explain the increased flux through this pathway, reveal differences in electrophoretic and kinetic behaviour, suggest a possible genetic basis for predisposition to this tumour, and provide an explanation in part for the therapeutic effects of ionizing radiation.

3. <u>Enzymes of Direct Oxidative Pathway</u>: It has long been recognized that the activity of this pathway is increased in many cancer tissues (14), an important functional objective being to provide increased pentoses for nucleic acid synthesis, and increased NADPH for anabolic reactions where this is the preferred or obligatory cofactor. In company with many other investigators, we had earlier demonstrated increased activity of 6-phosphogluconate dehydrogenase (6-PGDH) in samples of cervical cancer tissue (15). By then, many authors had suggested measurement of

6-PGDH activity in vaginal washings as a reliable test for invasive cancer of the cervix (16,17). In a detailed analysis, we showed that this test was more reliable if the extracellular fluid of the vagina exclusive of its cellular content was used for the test (18), and that under these circumstances, most patients with invasive cancer of the cervix could be reliably diagnosed and that little overlap occurred from patients with benign gynecologic lesions, although as with ribonuclease, the test was unreliable in detecting *in situ* lesions (5).

In our more extensive and recent studies, we have confirmed the increase in 6-PGDH occurring in cervical cancer and have shown that this is also accompanied by increased glucose-6-phosphate dehydrogenase activity (7). Electrophoretic studies have revealed an interesting feature occurring in a high proportion of samples from cervical cancer tissue where, as opposed to a single band of 6-PGDH activity detectable on starch gels, many tumours manifest a duplet band in the same region (Figure 5). The exact significance of this duplet is not known; conversion from the single to the double band can be accelerated during aging of the samples, but only in the case of malignant tissues, suggesting that this is certainly not a phenotypic phenomenon but a post-transcriptional feature which may have to do with the action of proteases and sialidases present in higher concentration in the malignant tissues (10).

ENZYMES IN DIAGNOSIS AND MONITORING OF CANCER

In the first part of this presentation, I have described, using the model of cancer of the uterine cervix for this purpose, how fundamental research into the enzymology of cancer can provide clues to basic mechanisms involved in neoplasia, and can also point the way to potentially useful diagnostic procedures, although few have survived critical analysis to the point where they have become mandatory in cancer diagnosis. Other diagnostic tests have become fashionable on the basis of empirical observations, waxing and waning as initially encouraging reports are followed by more reserved and even downright condemnatory appraisals. Many such tests have now been laid to rest and I propose to deal only with those in whom a semblance of living utility still breathes. Interestingly, the importance of some of these empirically-introduced tests and their relation to the fundamental biochemistry of neoplasia has only become apparent long after the test was established in the diagnostic laboratory.

Acid Phosphatase: The history of this, one of the oldest tests to be used in the diagnosis of cancer, needs no recapitulation since it was the subject of a scholarly review by Bodansky (19). At the beginning of this decade, the question, "how sensitive is the test in cancer detection?" could legitimately be answered, "no better than a finger in the rectum!" The related question, "how specific is the test for prostatic cancer?" could evoke some cynical comments about a raised level in the female calling for a differential diagnosis between breast and prostatic cancer! These jaundiced remarks are more applicable to acid phosphatase determination performed with non-discriminatory substrates. Attempts to develop substrates "specific" for the prostatic enzyme have been almost universally unsuccessful. The pioneer work of King and of Fishman showed that the erythrocyte enzyme could be inhibited by formaldehyde and the prostatic enzyme by L-tartrate (20,21). Although the tartrate-inhibited acid phosphatase for a while become synonymous with "prostatic acid phosphatase," it was soon realized that this is a feature of the classical lysosomal acid phosphatase present in virtually all cells of the body, and quite significant enzyme levels were detected in the serum and urine of females. Newer approaches relied on so-called "prostatic-specific substrates" such as α-napthyl phosphate, thymolphthalein monophosphate, and 3'-AMP. In no sense are these truly specific for the prostatic enzyme, but they have the merit of being resistant to hydrolysis by some acid phosphatases of other cell types which may be found in serum and to that extent are preferable to the older substrates phenyl phosphate and β-glycerol phosphate.

A few years ago we developed a simple colorimetric method to measure acid phosphatase activity using the substrate 3-AMP and evaluated its efficiency as a diagnostic test for prostatic cancer (22). Table 3 compares our results with those of earlier series in the literature and emphasizes the superiority of this procedure in our hands. It should be noted that we are comparing our work with older publications where the standards of technical performance may not have been as high as they are today. Moreover, ours is the smallest series and may have been more carefully analyzed through close clinical collaboration. In no sense would we imply that the test is "prostate specific" but we do feel that the earlier procedures have no place in diagnosis and that ours represents a test of minimum utility against which proposed improved assays can be judged.

Recent developments with acid phosphatase are worth a brief mention. Potentially the most exciting is the development of a radioimmunoassay procedure which is said to be specific for prostatic acid phosphatase. Some reports on its clinical application have appeared (23,24) and it appears to offer improved diagnostic specificity over total acid phosphatase and tartrate-sensitive acid phosphatase assays in the diagnosis of prostatic cancer. Table 4 summarizes some data from one of these papers and emphasizes the superiority of the RIA procedure over enzyme assay at each stage of the disease; but it also indicates that 6% of patients with benign hypertrophy had raised acid phosphatase detectable by RIA. The more sensitive and specific a test for prostatic cancer, the more important will it be to avoid rectal examination and defecation prior to blood sampling in order to secure reliable results since these have been a source of interference and misinterpretation in the past (25).

Electrophoresis has also been used to characterize acid phosphatase isoenzymes, and the system utilizing polyacrylamide gels as developed by Yam (26) seems to be the most discrete and informative. As indicated in Figure 6, aqueous extracts of prostate contain two isoenzymes designated 3 and 5; platelets contain two designated 2 and 5, and this is also the pattern of normal serum; isoenzymes 2 and 5 are greatly increased in patients with prostatic cancer whereas patients with breast cancer show increase predominantly in isoenzymes 3 and 5. Isoenzyme 5 is tartrate resistant (it is thus surprising that it should be increased in patients with prostatic cancer where increase in tartrate sensitive acid phosphatase has long been considered the dominant abnormality) and is specifically increased in patients with "hairy cell" leukemia. A tartrate resistant acid phosphatase is also increased in patients with Gaucher's disease and recent publications by Glew have emphasized the major differences between this isoenzyme and that normally raised in prostatic cancer (27,28).

Lactate Dehydrogenase (LDH): This enzyme has been used in the diagnosis of cancer for several decades, but appears to be increased only in advanced malignant disease, especially where hepatic metastases are present. Increase in urinary LDH occurs in renal tract cancers, but comparable increases occur in inflammatory disease of the renal tract so that the test lacks specificity and is hardly worth doing in suspected cases (29). We have compared the enzyme glutathione reductase with LDH and aspartate aminotransferase in the serum of a group of cancer patients, most of whom did not have metastatic disease. Of the three enzymes tested, glutathione reductase showed the highest incidence of abnormal values in this series (30) although activities did not seem to correlate with the extent of the disease. LDH was rather disappointing and does not seem to have a place in the diagnosis of non-metastatic cancer (Table 5). Occasionally, a variant LDH isoenzyme may be the first indication of cancer in a patient. An example is shown in Figure 7 taken from the work of Wilkinson (31); seven bands of LDH activity were noted in this patient's serum and subsequently explained by a mutation of the M-peptide of LDH in the tumour tissue which had a faster mobility than the normal M-polypeptide. As a consequence, the tumour isoenzymes which were released into the serum moved more rapidly than the normal isoenzymes although several tended to overlap with the latter so that only two extra bands rather than four extra bands could be detected.

It has long been believed that the LDH content of tumours undergoes transformation to an LD5-dominant pattern, irrespective of the pattern of the normal adult tissue. This has been interpreted as a reversion to a pattern typical of fetal tissues and better suited to anaerobic existence (32). This is certainly true of many tumours, and indeed such a transformation can often be seen as a pre-malignant change. However, in the cancers of the uterine cervix which we studied, changes in two directions from the symmetrical pattern encountered in normal cervix tissues were seen: shift to a more cathodal pattern said to be typical of cancers; and shift to a more anodal pattern which was rather unexpected and could not be explained by erythrocyte contamination of the samples (10). Although our study was not set up to systematically examine this possibility, it is feasible that those tumours showing the cathodal shift were deeper and more advanced to the point where the blood supply was being compromised, whereas those with the shift to the anodal distribution could have been more superficial or better endowed with blood vessels so that metabolism was largely aerobic.

Biliary Tract Enzymes: A variety of enzymes which are loosely described under this category have been introduced into cancer diagnosis and many still persist as routine procedures. Historically, alkaline phosphatase was the first, and it is still the most widely used. I do not proposed to describe its role in cancer diagnosis beyond commenting on its relative merits compared with newer tests, since Dr. Fishman has dealt with the more recent and exciting aspects of where precisely this enzyme fits into the cancer schema. Further consideration in this section will be restricted to 5'-nucleotidase (5NT) and γ-glutamyltransferase (GGT).

5NT has had a special interest for this reviewer. From the point of view of the neoplastic process itself, a number of observations have been made. These include the fact that tissues with high 5NT activity tend to have a low RNA content and low DNA turnover; regenerating tissues have low 5NT content as do the tissues of the newborn; reduced activity of this enzyme has been noted in the plasma membrane of rat hepatomas (33,34). The normal serum activity is much lower in infancy than in adults (35), and the 5NT content of bone tissue is much lower in infants than in adults (36). In view of the fact that various isoenzymes of 5NT appear to exist in various intracellular loci, and in some of these tissues reciprocity between 5NT and alkaline phosphatase exists in that adult femur has high 5NT and low alkaline phosphatase activity--the reverse of the situation in the infantile bone (Table 6), it is conceivable that in some tissues 5NT exerts its regulatory effect on growth by enhanced breakdown of nucleotide precursors. In rapidly growing tissues, suppression of 5NT activity will lead to conservation of nucleotides with enhanced alkaline phosphatase providing the required inorganic phosphate from non-nucleotide phosphate esters.

The sensitivity of 5NT in detecting the presence of liver metastases was compared to that of seven other serum enzyme tests in a large study involving the examination of nearly 1,000 cases with a suspected hepatobiliary disease (37). As shown in Table 7, 5NT was the enzyme most frequently raised in subjects with liver metastases reaching an incidence of 91%; in the absence of metastases, only 20% of the patients had raised serum 5NT activity. In another recent study involving over 70 patients with malignant lymphoma who, on first presentation, were carefully classified according to the extent of organ involvement, 20 patients were found to have clinical, biopsy, or scintigraphic evidence of lymphomatous infiltration of the liver. Of these, 14 patients demonstrated raised serum 5NT activity, a much higher incidence than that given by four other enzymes tested, including alkaline phosphatase, although the occasional patient with a normal serum 5NT activity and liver involvement demonstrated abnormality of one of the other enzymes (Table 8). It is our opinion that serum 5NT determination should be carried out in the initial assessment of all lymphoma patients (38).

Several studies by Dutch workers have emphasized the value of 5NT in monitoring patients during cancer therapy (39-40). The superiority of 5NT over alkaline phosphatase in this regard is shown in Figure 8, where successful estrogen therapy of a patient with breast cancer is followed by a prompt fall in the former enzyme whereas the latter continues to rise, possibly due to healing of bone secondaries. Figure 9 compares 5NT with GGT and ALT in monitoring the therapy of two patients; the first treated with hormones and the second with chemotherapy. Whereas GGT increased in both subjects, despite a clinically favorable course, due to the effects of therapy *per se* on the enzyme activity, 5NT remained essentially within normal limits in the first case and showed a steady decline correlating well with the clinical progress in the second. These authors have argued strongly for the value of serum 5NT determinations as an essential adjunct in the therapeutic monitoring of cancer patients.

The role of GGT in the detection of hepatic metastases has been controversial (41,42). The test is extremely sensitive and is raised to very high levels in most patients with liver metastases. The problem is that elevations of the same magnitude occur in other diseases of the biliary tree so that a decision point appropriate for hepatic metastases cannot be defined to the exclusion of other disease states, particularly those with biliary tract involvement where high serum levels of GGT are encountered. Very high activities are also seen in alcoholic liver disease and cirrhosis. Even alcohol abuse without liver disease, and ingestion of many drugs are now recognized to cause quite significant increases in serum GGT activity. A special situation recently reported calls attention to the high GGT content of ascitic fluid from patients with the accumulation of fluid in the peritoneal cavity due to hepatomas as opposed to cirrhosis or liver metastases (43). This enzyme is undoubtedly causing enormous interest in experimental carcinogenesis. Very high levels are found in chemically-induced hepatomas in experimental animals (44) and recent work suggests that the enzyme may be used as a marker of pre-neoplastic cells induced by carcinogens (45). The implications of this discovery are very important, since the increase appears to be approximately two orders of magnitude, and should enable the life history of the hepatoma cell to be followed from its earliest stage when it becomes committed to neoplasia until the final transformation is evident.

Ribonuclease: Early investigations on the clinical usefulness of this enzyme in cancer diagnosis led to disappointing results (46,47). Its activity was increased in the urine (48) but not in the serum (49) of patients with cancer of the cervix undergoing therapeutic radiation. More recent reports suggested that it might be useful in certain forms of neoplasia, multiple myeloma being particularly favored (50,51). However, careful studies have now shown that impaired renal function and poor nutrition are the main factors leading to such abnormalities in cancer patients, and the malignant process itself plays a very secondary role (52-54). One recent paper describes a special use for the enzyme in the diagnosis of pancreatic cancer (55). As shown in Figure 10, values above 300 Units were encountered only in such patients (75% of the 30 studied) and never in 52 normal subjects, 10 with pancreatitis, or 69 with other forms of cancer. While this incidence of 75% for raised serum ribonuclease activity in pancreatic cancer has been confirmed (56), the authors found comparable increases in patients with other cancers and even those with non-cancerous disease. The original specificity claimed for this test does not therefore seem to be justified. Ovarian cancer has been reported as another special tumor form in which serum ribonuclease has a valuable role in diagnosis and staging with increased activities found in more than 90% of patients in one series (57) but this claim awaits confirmation.

Lysozyme: Activity of this enzyme varies in the different leukemias, and measurements have been made on both serum and urine.

Lysozyme activity is reduced in both sources in acute lymphatic leukemia, but considerably raised above normal in the other forms of this disease, and most of all in the acute monocytic form (58, 59). It has been recommended as having a special role in monitoring the progress of children with acute leukemia (60). Correlation exists between the lysozyme activity of serum and of the peripheral white blood cells in these conditions (61). A striking difference is found in the ratio of the urinary to serum enzyme, and this may be a further aid in diagnosis; this ratio is 10-fold higher in patients with acute myelomonocytic leukemia compared to subjects with chronic myeloid leukemia (59). Proliferation of macrophages may be responsible for increased lysozyme production which may accumulate in the renal tubules and thus contribute to the increased activity of the enzyme in the urine with impaired tubular reabsorption of the filtered lysozyme contributing further to this increase (62). Although elevated lysozyme activities occur in serum and urine in a variety of renal diseases, in lymphoma, and in disseminated cancer, these are less dramatic than the increases found in the myeloid and monocytic leukemias; the increases in many cancer patients, and in particular those encountered in multiple myeloma, appear to reflect impaired renal function rather than the malignant process *per se* (63). The role of the lysozyme in diagnosing patients with colonic cancer is marred by the increases which also occur in inflammatory bowel disease (64), but it may have a place in monitoring the former patients since previously elevated serum activity fell to normal on excision of the tumor in half the subjects studied by one group of authors (65).

Other Enzymes: Among the newer enzymes on which reports have appeared in the last year or two, and which await more complete evaluation, mention should be made of the fucosyltransferases which are related to the topic of Dr. Weiser's presentation (66,67). Many routinely established enzyme tests have been applied to cancer diagnosis even though their primary diagnostic function relates to other fields such as hepatobiliary disease and myocardial infarction. Bodansky (68) has assembled data on many of these nonspecific ubiquitous enzymes which are present in all body cells, and on the basis of his evaluation of their utility in cancer diagnosis, he concluded that phosphohexose isomerase was the most generally useful (Table 9). Unlike many of the other enzymes listed, phosphohexose isomerase is not simply a marker for hepatic involvement. Very high increases can be encountered in cancers of the breast and prostate correlating with the degree of bone metastases (69,70).

ENZYMES IN SOURCES OTHER THAN SERUM

Space does not allow full coverage of this extensive topic. A few examples will serve to illustrate the diversity of approaches to the diagnosis of cancer and the reader will be referred to a few pertinent reports and reviews.

Urine Enzymes: This topic has been comprehensively discussed by Schwartz (71) and by Raab (72). Urinary enzymes have not proven of durable value in cancer diagnosis although publications on this topic continue to appear. Recently, attention was drawn to the use of urinary excretion of arylsulfatase A in detecting patients with transitional cell carcinoma of the bladder (73) and Figure 11 illustrates data comparing this enzyme with three others. Unfortunately, the numbers were not the same for each enzyme and confirmation of this work will be required. The report stated that normal values for all four enzymes occurred in a male whose cancer had been surgically removed.

Ascitic Fluid: As previously mentioned, γ-glutamyl transferase has been measured in this source and much higher activities occurred in cases of primary hepatoma than in patients with secondary carcinoma whose activities overlapped those found in patients with non-malignant conditions (43).

Bone Marrow: Total acid phosphatase determinations have been performed in this material and it has been claimed that increased bone marrow activity can be detected in the presence of metastases when the serum enzyme remains normal and is therefore a useful procedure in staging patients with prostatic cancer (74, 75). A more recent study reported unfavorably on the use of total acid phosphatase determinations in bone marrow aspirate (76). Use of the new radioimmunoassay test for prostate-specific acid phosphatase is likely to be a more rewarding tool as applied to bone marrow aspirates (24).

Vaginal Fluid: This source has been used as a means of diagnosing gynecologic cancers by enzyme determinations, among which β-glucuronidase, 6-phosphogluconate dehydrogenase, and ribonuclease have been most widely tested (8,71,77,78). These procedures have not entered into routine use although they have considerable potential in the economical screening of patients with invasive cervical cancer.

Duodenal Aspirate: Deficiency of the exocrine pancreas due to cancer is best detected by measuring the output of pancreatic enzymes following appropriate stimulation (79,80). Reduction in the output of trypsin and chymotrypsin may be detectable even in patients who are neither icteric or suffering from steatorrhoea and whose cancer is at a stage when therapy can be contemplated (81).

DAVID GOLDBERG

CONCLUSION

As with other aspects of laboratory medicine, reliable diagnostic tests will only become available when we are more certain of the etiology of the disease and are in possession of more fundamental knowledge about the enzymology of human cancer. Many empirical observations have led to the plethora of existing tests which need to be streamlined so that those which are ineffectual are weeded out and those which hold promise are developed to the highest technical standards with emphasis on cost-effectiveness. Whereas individual enzyme tests have often proven disappointing in the light of initial enthusiastic advocacy, combinations of such tests, and of enzyme tests together with hormone and protein tests described in this Symposium, may help in developing cancer-specific profiles from tests which in themselves lack adequate specificity. As the number of tests used in such profiles grow, it will be necessary to use mathematical techniques such as discriminant function analysis to aid in their interpretation. This type of procedure holds great potential and has already been shown to be quite effective in discriminating malignant from non-malignant lesions of the liver and biliary tree (82).

REFERENCES

1. Goldberg, D.M., Pitts, J.F.: Enzymes of the human cervix uteri. Comparison of nucleases and adenosine deaminase in malignant and non-malignant tissue samples. Br. J. Cancer 20: 729-742, 1966.
2. Goldberg, D.M., Ayre, H.A., Pitts, J.F.: Effect of radium treatment on activity and distribution of some enzymes in cancers of the human cervix uteri. Cancer 20:1388-1394, 1967.
3. Goldberg, D.M.: Alkaline ribonuclease activity in response to therapeutic radiation in the human female, in Biochemical Indicators of Radiation Injury in Man. International Atomic Energy Agency, Vienna, pp 259-275, 1971.
4. Goldberg, D.M., Hart, D.M., Watts, C.: Vaginal fluid ribonuclease activity in the diagnosis of cervical cancer. J. Obstet. Gynaecol. Br. Cwlth. 75:762-767, 1968.
5. Goldberg, D.M., Watts, C., Hart, D.M.: Evaluation of several enzyme tests in vaginal fluid as aids to the diagnosis of invasive and preinvasive cervical cancer. Amer. J. Obstet. Gynecol. 107:465-471, 1970.
6. Warburg, O.: On the origin of cancer cells. Science 123: 309-314, 1956.
7. Marshall, M.J., Goldberg, D.M., Neal, F.E., Millar, D.R.: Enzymes of glucose metabolism in carcinoma of the cervix and endometrium of the human uterus. Br. J. Cancer. 37:990-1001, 1978.
8. Goldberg, D.M.: Enzymology of uterine cancer, in 6th International Symposium on Clinical Enzymology. Eds. A. Burlina, L. Galzigna. Kurtis, Milan, pp. 235-253, 1976.
9. Kikuchi, Y., Sato, S., Sugimura, T.: Hexokinase isoenzyme patterns of human uterine tumors. Cancer 30:444-447, 1972.
10. Marshall, M.J., Goldberg, D.M., Neal, F.E.: Changes in isoenzyme composition of some enzymes of carbohydrate metabolism in cancer of the human cervix uteri. In press.
11. Spencer, N., Hopkinson, D.A., Harris, H.: Phosphoglucomutase polymorphism in man. Nature 204:742-745, 1964.
12. Marshall, M.J., Goldberg, D.M., Neal, F.E., Millar, D.R.: Properties of glycolytic and related enzymes of normal and malignant human uterine tissues studied to optimise assay conditions. Enzyme 23:295-306, 1978.
13. Marshall, M.J., Neal, F.E., Goldberg, D.M.: The effect of radiotherapy upon enzymes of the glycolytic and related pathways in human uterine cancer. Br. J. Cancer, in press.
14. Weber, G.: Enzymology of cancer cells. NEJM 296:541-551,1977.
15. Ayre, H.A., Goldberg, D.M.: Enzymes of the human cervix uteri. Comparison of dehydrogenases of lactate, isocitrate and phosphogluconate in malignant and non-malignant tissue samples. Br. J. Cancer 20:743-750, 1966.
16. Bonham, D.G., Gibbs, D.F.: A new test for gynaecological

cancer--6-phosphogluconate dehydrogenase activity in vaginal fluid. Br. Med. J. 2:823-824, 1962.
17. Bell, J.L., Egerton, M.E.: 6-Phosphogluconate dehydrogenase estimation in vaginal fluid in the diagnosis of cervical cancer. J. Obstet. Gynaecol. Br. Cwlth. 72:603-609, 1965.
18. Goldberg, D.M., Watts, C., Hart, D.M.: Diagnostic test for cervical cancer. Br. Med. J. 1:424-425, 1968.
19. Bodansky, O.: Acid phosphatase. Adv. Clin. Chem. 15:43-147, 1972.
20. Abul-Fadl, M.A.M., King, E.J.: Properties of the acid phosphatases of erythrocytes and of the human prostate gland. Biochem. J. 45:51-60, 1949.
21. Fishman, W.H., Lerner, F.: A method for estimating serum acid phosphatase of prostatic origin. J. Biol. Chem. 200: 89-97, 1953.
22. Goldberg, D.M., Ellis, G.: An assessment of serum acid and alkaline phosphatase determinations in prostatic cancer with a clinical validation of an acid phosphatase assay utilizing adenosine 3'-monophosphate as substrate. J. Clin. Pathol. 27:140-147, 1974.
23. Foti, A.G., Cooper, J.F., Herschman, H., Malvaez, R.R.: Detection of prostatic cancer by solid-phase radioimmunoassay of serum prostatic acid phosphatase. NEJM 297:1357-1361, 1977.
24. Belville, W.D., Cox, H.D., Mahan, D.E., Olmert, J.P., Bernhard, T.M., Bruce, A.W.: Bone marrow acid phosphatase by radioimmunoassay. Cancer 41:2286-2291, 1978.
25. Daniel, O., Van Zyl, J.J., Rise of serum-acid-phosphatase level following palpation of the prostate. Lancet 1:998-999, 1952.
26. Yam, L.T.: Clinical significance of the human acid phosphatases. Amer. J. Med. 56:604-616, 1974.
27. Mercer, D.W., Peters, S.P., Glew, R.H., Lee, R.E., Wenger, D.M.: Acid phosphatase isoenzymes in Gaucher's disease. Clin. Chem. 23:631-635, 1977.
28. Chambers, J.P., Peters, S.P., Glew, R.H., Lee, R.E., McCafferty, L.R., Mercer, D.W., Wenger, D.A.: Multiple forms of acid phosphatase activity in Gaucher's disease. Metabolism 27:801-814, 1978.
29. Wacker, W.E.C., Dorfman, L.E.: Urinary lactic dehydrogenase activity: 1. Screening method for detection of cancer of kidneys and bladder. J. Amer. Med. Assoc. 181:972-978, 1962.
30. Delides, A., Spooner, R.J., Goldberg, D.M., Neal, F.E.: An optimized semi-automatic rate method for serum glutathione reductase activity and its application to patients with malignant disease. J. Clin. Pathol. 29:73-77, 1976.
31. Wilkinson, J.H.: Clinical applications of isoenzymes. Clin. Chem. 16:733-739, 1970.
32. Latner, A.L., Skillen, A.W.: Isoenzymes in Biology and Medicine. Academic Press, New York, 1968.

33. Goldberg, D.M.: 5'-nucleotidase: recent advances in cell biology, methodology, and clinical significance. Digestion 8:87-99, 1973.
34. Goldberg, D.M.: Biochemical and clinical aspects of 5'-nucleotidase in gastroenterology, in Frontiers of Gastrointestinal Research. Vol. 2., Ed. L. Van der Reis, Karger, Basel, pp 71-108, 1976.
35. Belfield, A., Goldberg, D.M.: Normal ranges and diagnostic value of serum 5'-nucleotidase and alkaline phosphatase activities in infancy. Arch. Dis. Child. 46:842-846, 1971.
36. Goldberg, D.M., Belfield, A.: Reciprocal relationship of alkaline phosphatase and 5'-nucleotidase in human bone. Nature 247:286-288, 1974.
37. Ellis, G., Goldberg, D.M., Spooner, R.J., Ward, A.M.: Serum enzyme tests in diseases of the liver and biliary tree. Amer. J. Clin. Pathol. 70:248-258, 1978.
38. Deeble, T.J., Goldberg, D.M.: Biochemical tests for bone and liver involvement in malignant lymphoma patients. In press.
39. Van der Slik, W., Persijn, J-P., Engelsman, E., Riethorst, A.: Serum 5'-nucleotidase. Clin. Biochem. 3:59-80, 1970.
40. Korsten, C.B., Persijn, J-P., Van der Slik, W.: The application of the serum γ-glutamyl transpeptidase and the 5'-nucleotidase assay in cancer patients: a comparative study. Z. Klin. Chem. Klin. Biochem. 12:116-120, 1974.
41. Rosalki, S.B.: γ-glutamyl transpeptidase. Adv. Clin. Chem. 17:53-107, 1975.
42. Goldberg, D.M.: Structural, functional and clinical aspects of γ-glutamyl-transferase. CRC Crit. Revs. Clin. Lab. Sci. In press.
43. Peters, T.J., Seymour, C.A., Wells, G., Fakunle, F., Neale, G.: γ-glutamyltransferase levels in ascitic fluid and liver tissue from patients with primary hepatoma. Br. Med. J. 1:1576, 1977.
44. Fiala, S., Mohindru, A., Kettering, W.G., Fiala, A.E., Morris, H.P.: Glutathione and gamma glutamyl transpeptidase in rat liver during chemical carcinogenesis. J. Nat. Cancer Inst. 57:591-598, 1976.
45. Cameron, R., Kellen, J., Kolin, A., Malkin, A., Farber, E.: γ-glutamyltransferase in putative premalignant liver cell populations during hepatocarcinogenesis. Cancer Res. 38:823-829, 1978.
46. Levy, A.L., Rottino, A.: Effect of disease states on the ribonuclease concentration of body fluids. Clin. Chem. 6:43-51, 1960.
47. Zytko, J., Cantero, A.: Serum ribonuclease in patients with malignant disease. Can. Med. Assoc. J. 86:482-485, 1962.
48. Goldberg, D.M., MacVicar, J., Watts, C.: The effect of therapeutic radiation upon urinary excretion of nucleases and deoxyribosides in human patients with carcinoma of the cervix

uteri. Clin. Sci. 32:69-82, 1967.
49. Watts, C., Goldberg, D.M.: Serum enzymes after therapeutic radiation. A study in patients with carcinoma of cervix uteri. Enzymol. Biol. Clin. 8:379-386, 1967.
50. Fink, K., Adams, W.S., Skoog, W.A.: Serum ribonuclease in multiple myeloma. Amer. J. Med. 50:450-457, 1971.
51. Chretien, P.B., Matthews, W., Woomey, P.L.: Serum ribonuclease in cancer: relation to tumor histology. Cancer 31: 175-179, 1973.
52. Shenkin, A., Citrin, D.L., Rowan, R.M.: An assessment of the clinical usefulness of plasma ribonuclease assays. Clin. Chim. Acta 72:223-231, 1976.
53. Prabhavathi, P., Mohanram, M., Reddy, V.: Ribonuclease activity in plasma and leucocytes of malnourished children. Clin. Chim. Acta 79:591-593, 1977.
54. Karpetsky, T.P., Humphrey, R.L., Levy, C.C.: Influence of renal insufficiency on levels of serum ribonuclease in patients with multiple myeloma. J. Nat. Cancer Inst. 58:875-880, 1977.
55. Reddi, K.K., Holland, J.F.: Elevated serum ribonuclease in patients with pancreatic cancer. Proc. Nat. Acad. Sci. 73: 2308-2310, 1976.
56. Fitzgerald, P.J., Fortner, J.G., Watson, R.C., Schwartz, M.K., Sherlock, P., Benua, R.S., Cubilla, A.L., Schottenfeld, D., Miller, D., Winawer, S.J., Lightdale, C.J., Leidner, S.D., Nisselbaum, J.S., Menendez-Botet, C.J., Poleski, M.H.: The value of diagnostic aids in detecting pancreas cancer. Cancer 41:868-879, 1978.
57. Sheid, B., Lu, T., Pedrinan, L., Nelson, J.H., Jr.: Plasma ribonuclease. A marker for the detection of ovarian cancer. Cancer 39:2204-2208, 1977.
58. Perillie, P.E., Kaplan, S.S., Lefkowitz, E., Rogaway, W., Finch, S.C.: Studies of muramidase (lysozyme) in leukemia. J. Amer. Med. Assoc. 203:317-322, 1968.
59. Wiernik, P.K., Serpick, A.A.: Clinical significance of serum and urinary muramidase activity in leukemia and other hematologic malignancies. Amer. J. Med. 46:330-343, 1969.
60. Bratlid, D., Moe, P.J.: Serum lysozyme activity in children with acute leukemia. Eur. J. Ped. 127-263-268, 1978.
61. Noble, R.E., Fudenberg, H.H.: Leukocyte lysozyme activity in myelocytic leukemia. Blood 30:465-473, 1967.
62. Muggia, F.M., Heinemann, H.O., Farhangi, M., Osserman, E.P.: Lysozymuria and renal tubular dysfunction in monocytic and myelomonocytic leukemia. Amer. J. Med. 47:351-366, 1969.
63. Deo, E.A., Brook, J.: Serum lysozyme in multiple myeloma. J. Lab. Clin. Med. 90:899-903, 1977.
64. Pruzanski, W., Marcon, N., Ottaway, C., Prokipchuk, E.: Muramidase (lysozyme) in Crohn's disease and in ulcerative

colitis. Amer. J. Dig. Dis. 22:995-998, 1977.
65. Cooper, E.H., Turner, R., Steele, L., Goligher, J.C.: Blood muramidase activity in colo-rectal cancer. Br. Med. J. 3: 662-664, 1974.
66. Bauer, C., Köttgen, E., Reutter, W.: Elevated activities of α-2- and α-3-fucosyltransferases in human serum as a new indicator of malignancy. Biochem. Biophys. Res. Comm. 76:488-494, 1977.
67. Khilanani, P., Chou, T.H., Ratanatharathorn, V., Kessel, D.: Evaluation of two plasma fucosyltransferases as marker enzymes in non-Hodgkin's lymphoma. Cancer 41:701-705, 1978.
68. Bodansky, O.: Biochemistry of Human Cancer, Academic Press, New York, 1975.
69. Bodansky, O.: Serum phosphohexose isomerase in cancer: as index of tumor growth in metastatic carcinoma of breast. Cancer 7:1200-1226, 1954.
70. Bodansky, O.: Serum phosphohexose isomerase in cancer. III. As an index of tumor growth in metastatic carcinoma of the prostate. Cancer 8:1087-1114, 1955.
71. Schwartz, M.K.: Enzymes in cancer. Clin. Chem. 19:10-22, 1973.
72. Raab, W.: Diagnostic value of urinary enzyme determination. Clin. Chem. 18:5-25, 1972.
73. Posey, L.E., Morgan, L.R.: Urine enzyme activities in patients with transitional cell carcinoma of the bladder. Clin. Chim. Acta 74:7-10, 1977.
74. Chua, D.T., Veenema, R.J., Muggia, F., Graff, A.: Acid phosphatase levels in bone marrow: value in detecting early bone metastasis from carcinoma of the prostate. J. Urol. 103:462-466, 1970.
75. Reynolds, R.D., Greenberg, B.R., Martin, N.D., Lucas, R.N., Gaffney, C.N., Hawn, L.: Usefulness of bone marrow acid phosphatase in staging carcinoma of the prostate. Cancer 32:181-184, 1973.
76. Boehme, W.M., Augspurger, R.R., Wallner, S.F., Donohue, R.E.: Lack of usefulness of bone marrow enzymes and calcium in staging patients with prostatic cancer. Cancer 41:1433-1439, 1978.
77. Goldberg, D.M.: Clinical Enzymology, in Progress in Medicinal Chemistry, Vol. 13., Eds. G.P. Ellis and G.B. West. North Holland, Amsterdam. pp. 1-158, 1976.
78. Schwartz, M.K.: Laboratory aids to diagnosis - enzymes. Cancer 37:542-548, 1976.
79. Howat, H.T., The Pancreas, in Biochemical Disorders in Human Disease, 3rd. ed. Eds. R.H.S. Thompson and I.D.P. Wooton. Academic Press, New York. pp 703-722, 1970.
80. Gowenlock, A.H.: Diseases of the Alimentary Tract, in The Principles and Practice of Diagnostic Enzymology. Ed. J. H. Wilkinson. Arnold, London. pp. 361-398, 1976.

81. Sale, J.K., Goldberg, D.M., Fawcett, A.N., Wormsley, K.G.: Trypsin and chymotrypsin as aids in the diagnosis of pancreatic disease. Amer. J. Dig. Dis. 17:780-792, 1972.
82. Goldberg, D.M., Ellis, G.: Mathematical and computer-assisted procedures in the diagnosis of liver and biliary tract disorders. Adv. Clin. Chem. 20:49-128, 1978.

TABLE 1. ENZYME ACTIVITIES INVOLVED IN NUCLEIC ACID BREAKDOWN IN SUPERNATANT FRACTION OF NORMAL AND MALIGNANT CERVIX UTERI (from Ref. 1)

	Units/mg. Protein			
	Normal	Cancer	t	P
Alkaline Ribonuclease	45.6 ± 6.9	532.4 ± 80.7	4.77	<0.001
Acid Ribonuclease	19.3 ± 3.4	298.3 ± 40.9	6.86	<0.001
Deoxyribonuclease I	0.42 ± 0.15	1.21 ± 0.25	2.33	<0.05
Deoxyribonuclease II	1.91 ± 0.30	8.58 ± 1.01	5.16	<0.001
Adenosine Deaminase	400 ± 42	1204 ± 210	2.99	<0.005

Mean ± S.E. of 16 normals and 25 cancers. Ribonucleases as µg. RNA-P solubilised/hour/mg. protein and deoxyribonucleases as µg. DNA-P solubilised/hour/mg. protein. Adenosine Deaminase as µM deaminated/hour/mg. protein.

TABLE 2. SPECIFIC ACTIVITIES (U/g protein) OF ENZYMES OF GLYCOLYTIC AND DIRECT OXIDATIVE GLUCOSE PATHWAYS IN NORMAL AND MALIGNANT SAMPLES OF THE CERVIX AND ENDOMETRIUM OF THE UTERUS. Data as Mean (s.d.) From Ref. 7

	CERVIX		ENDOMETRIUM	
	Normal	Malignant	Normal	Malignant
Glucose-6-phosphate dehydrogenase	27.8 (13.3)	65.0 (49.7)	44.7 (14.2)	48.9 (41.7)
6-Phosphogluconate dehydrogenase	22.2 (12.6)	56.3 (37.0)	35.5 (11.6)	49.9 (30.9)
Hexokinase	18.8 (5.34)	30.3 (19.6)	28.9 (10.3)	28.1 (16.1)
Phosphoglucose Isomerase	608 (161)	1660 (1260)	983 (303)	1760 (1200)
Phosphoglucomutase	195 (53.5)	249 (131)	328 (141)	338 (198)
Phosphofructokinase	45.1 (19.1)	71.5 (45.5)	97.3 (37.8)	91.4 (61.4)
Aldolase	16.7 (10.1)	62.4 (39.1)	43.3 (16.2)	55.7 (36.5)
α-Glycerol phosphate dehydrogenase	6.00 (1.87)	9.12 (10.3)	11.2 (7.16)	11.4 (7.82)
Glyceraldehyde 3-phosphate dehydrogenase	229 (63)	586 (332)	548 (173)	570 (299)
Enolase	283 (65)	739 (386)	538 (170)	686 (348)
Pyruvate kinase	503 (207)	1910 (2550)	582 (192)	1440 (1040)
Lactate Dehydrogenase	903 (212)	2290 (1190)	2260 (859)	2800 (1420)

TABLE 3. PERCENTAGE INCIDENCE OF ACID PHOSPHATASE ELEVATION IN PROSTATIC CANCER

Reference[a]	Sullivan et al. [148a]	Herbert [148b]	Woodard [148c]	Marshall and Amador [143]	Murphy et al. [141]	Goldberg and Ellis [148]
Total number of cases	200	82	178	107	185	46
Substrate	Phenyl phosphate			β-Glycerophosphate		3' AMP
Stages:						
A	⎱ 11	⎱ 42	⎱ 25	⎱ 46	45	25
B	⎰	⎰	⎰	⎰	36	80
C			60	56	57	100
D	85	89	74	83	88	100
All Stages	59	62	65	56	64	83

[a]The full references are given in the original paper.[22]

TABLE 4. COMPARISON OF PROSTATIC ACID PHOSPHATASE BY RADIOIMMUNO-ASSAY (RIA) AND TARTRATE-SENSITIVE ENZYME ASSAY
From Ref. 23

Group	No.	Raised by RIA (%)	Raised by Enzyme Assay (%)
Prostatic Cancer:			
Stage 1	24	33	12
2	33	79	15
3	31	71	29
4	25	92	60
Benign Hypertrophy	36	6	0
Other Cancers	83	11	8

TABLE 5. RELATION BETWEEN ABNORMAL ENZYMES AND SURVIVAL IN CANCER PATIENTS AFTER SIX MONTHS FROM TIME OF EXAMINATION
From Ref. 30

	Percentage with Raised Serum Enzymes[a]			
	LDH	GR	AST	ALP
Dead (36)[b]	48	57	20	29
Alive (32)[b]	18	47	0	9

[a] Abbreviations as follows: LDH, lactate dehydrogenase; GR, glutathione reductase; AST, aspartate aminotransferase; ALP, alkaline phosphatase.

[b] Number of cases in parentheses

TABLE 6. SPECIFIC ACTIVITIES OF NUCLEOTIDE HYDROLYSIS OF VARIOUS CATEGORIES OF ADULT AND NEONATAL BONE AS mU PER MG PROTEIN NITROGEN. (From Reference 36)

Tissue	No. of Specimens	2'-AMPase*	3'-AMPase*	5'AMPase*
Shaft of femur (children)	5	174.2 ± 36.5	162.2 ± 30.4	18.5 ± 1.6
Shaft of femur (adults)	6	18.0 ± 5.5	25.0 ± 4.4	73.1 ± 24.1
Paget's disease (femur)	5	167.4 ± 41.1	159.3 ± 40.7	46.1 ± 8.2

*Mean ± s.e.

Note: Hydrolysis of 2' AMP and 3'-AMP is due to alkaline phosphatase whereas that of 5'-AMP is due to 5'-nucleotidase. The activity of the latter is low and of the former high in children where as in adults the reverse applies. In Paget's disease, the behaviour of these enzymes tends toward that seen in infant bone.

TABLE 7. PERCENTAGE ABNORMALITIES OF SOME SERUM ENZYME ACTIVITIES IN CANCER PATIENTS WITH AND WITHOUT HEPATIC METASTASES.
From Ref. 37

	% Incidence of Abnormal Activities	
	Liver Metastases Present	Liver Metastases Absent
Aspartate Aminotransferase	86	8
Alanine Aminotransferase	74	20
Isocitrate Dehydrogenase	76	27
Glutamate Dehydrogenase	81	16
Guanase	81	27
Adenosine Deaminase	41	19
5'-Nucleotidase	91	20
Alkaline Phosphatase	85	27

TABLE 8. ANALYSIS OF TWENTY LYMPHOMA PATIENTS WITH EVIDENCE OF HEPATIC INVOLVEMENT INITIALLY OR DURING FOLLOWUP
From Ref. 37

	Number with Abnormal Activities	
	Pts. with initially raised serum 5NT (n=14)	Pts. with initially normal serum 5NT (n=6)
Other Enzymes Elevated (No.)		
Alkaline Phosphatase	9	0
Aspartate Aminotransferase	7	2
Alanine Aminotransferase	6	1
Ornithine Carbamoyl Transferase	7	1

TABLE 9. PERCENTAGE INCIDENCE OF ABNORMAL VALUES FOR SOME COMMONLY DETERMINED ENZYMES IN CANCER PATIENTS. Based on Data collected by Bodansky (Ref. 68).

Type of Cancer	No. of Patients	PHI	ALD	ICD	MD	AST	ALT
Gastrointestinal	119	74	70	40	37	35	19
Head and Neck	88	51	45	21	10	16	9
Lung	126	72	62	24	25	15	7
Breast	57	70	45	40	40	-	-
Liver Metastases	284	84	75	53	62	51	44

Abbreviations: PHI, Phosphohexose Isomerase; ALD, Aldolase; ICD, Isocitrate Dehydrogenase; MD, Malate Dehydrogenase; AST, Aspartate Aminotransferase; ALT, Alanine Aminotransferase.

Figure 1. Change in vaginal fluid cell ribonuclease activity following radiotherapy in two patients with cancer of the uterine cervix. From Reference 3.

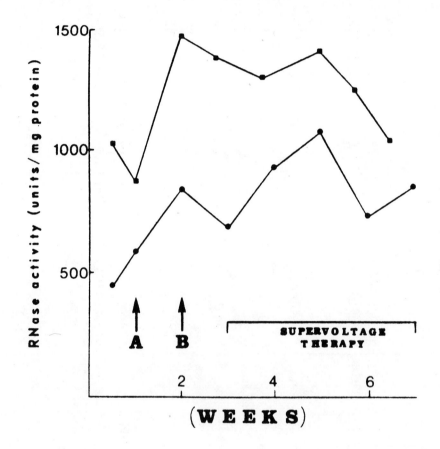

Figure 2. Relationship between phosphoenolpyruvate (PEP) concentration and pyruvate kinase activity at saturating ADP concentration in presence and absence of 0.3 mM fructose-1,6-diphosphate (FDP) in normal cervix. From Reference 12.

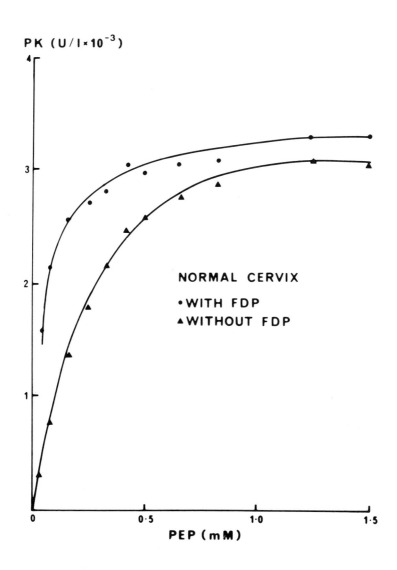

Figure 3: Relationship between phosphoenolpyruvate (PEP) concentration and pyruvate kinase activity at saturating ADP concentration in presence and absence of 0.3 mM fructose-1,6-diphosphate (FDP) in cancer of the cervix. From Reference 12.

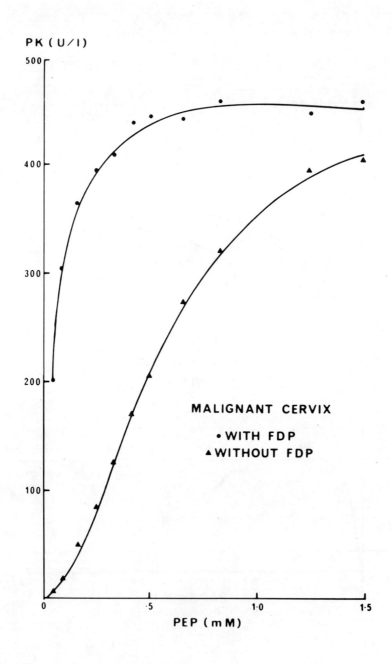

Figure 4. Phosphofructokinase activity in paired biopsy samples from cervical cancer patients before and after radiotherapy. From Reference 13.

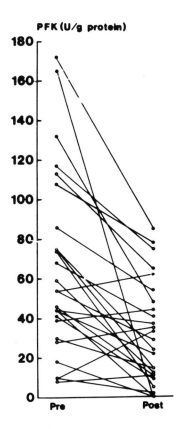

Figure 5. Zymograms of 6-phosphogluconate dehydrogenase in cervical cancer extracts. Specimens 1 and 4 show duplet form strongly and specimens 2 and 3 (numbered from left to right) show weaker duplet band. From Reference 10.

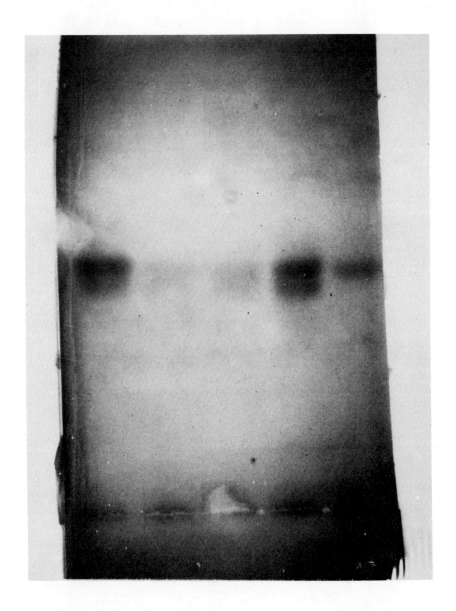

Figure 6. Acrylamide gel electrophoresis of extracts stained to show bands 2, 3 and 5 of acid phosphatase. Left to right: A, aqueous extract of prostate; B, aqueous extract of platelets; C, normal serum; D, serum from case of prostatic cancer; E, serum from patient with metastatic breast cancer. From reference 26, with kind permission of the author and publishers.

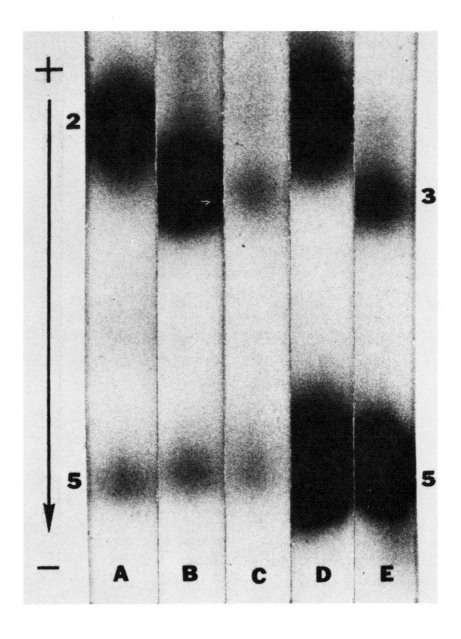

Figure 7. Top zymogram shows extra bands of LDH activity in serum of cancer patient. Middle and lower zymograms show LDH isoenzyme pattern of normal and cancerous tissue from the same patient. From Reference 31 with kind permission of the publishers.

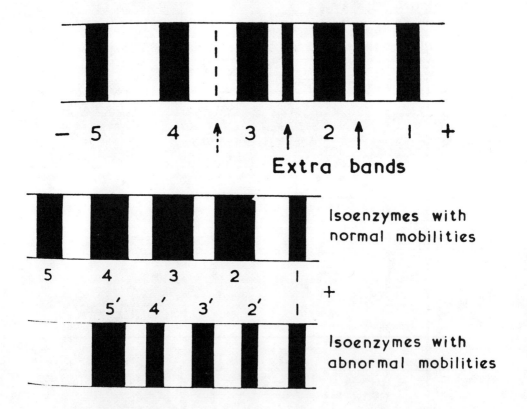

Figure 8. Variation with time, of 5'-nucleotidase (solid circles) and alkaline phosphatase (open circles), haemoglobin (Hb) and sedimentation rate (SRE) in a patient treated with oestrogen therapy as indicated by arrow on X-axis which represents time in months. The lower dotted line represents upper normal limit for alkaline phosphatase and 5'-nucleotidase (left and right of ordinate, respectively), and the upper dotted line represents mean normal Hb concentration. From Reference 39 with kind permission of the authors and publishers.

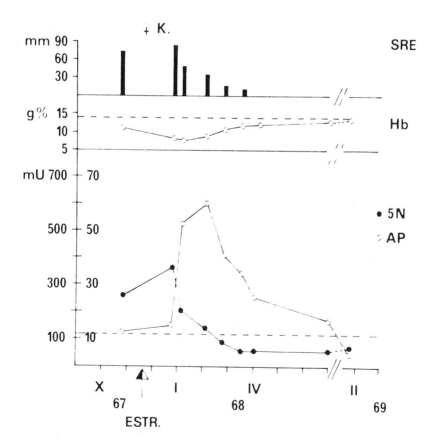

Figure 9. Variation, with time, of serum gamma glutamyl transferase activity (open circles), 5'-nucleotidase activity (solid circles) and alanine aminotransferase activity (triangles) in: a) patient with breast cancer treated by hormonal manipulation; b) patient with lymphoma treated with chemotherapy. From Reference 40, with kind permission of the authors and publishers.

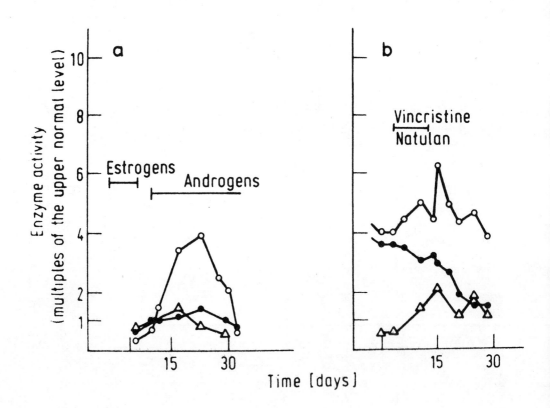

Figure 10. Serum ribonuclease activity plotted as cumulative percent of all persons in the following groups: normal, n = 52 (-); pancreatitis, n = 10 (- - -); varied cancers, n = 69 (...); pancreatic cancer, n = 30 (-.-.). From Reference 55, with kind permission of the authors and publishers.

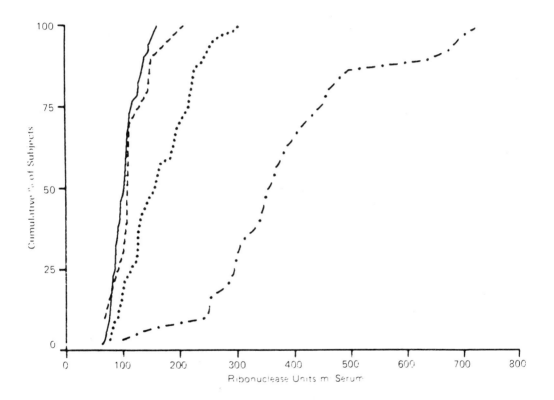

Figure 11. Urine enzyme activities in patients with active bladder cancer. Shaded areas normal range for each enzyme. From Reference 73, with kind permission of the authors and publishers.

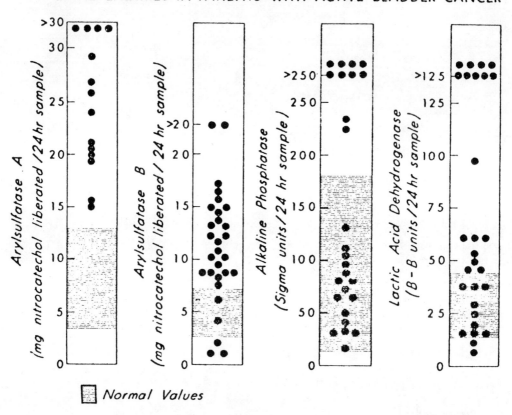

ACKNOWLEDGEMENT

The author is indebted to the following journals and publishers for permission to reproduce figures as follows: Figure 1, International Atomic Energy Agency; Figures 2 and 3, Enzyme; Figure 4, British Journal of Cancer; Figure 6, American Journal of Medicine; Figure 7, Clinical Chemistry; Figure 8, Clinical Biochemistry; Figure 9, Journal of Clinical Chemistry and Clinical Biochemistry; Figure 10, Clinica Chimica Acta.

CANCER-ASSOCIATED GALACTOSYLTRANSFERASE AND GLYCOPEPTIDE ACCEPTOR ACTIVITIES

Milton M. Weiser, M.D.[1] and Daniel K. Podolsky, M.D.[2]

[1]*Chief, Division of Gastroenterology and Nutrition, Department of Medicine, State University of New York at Buffalo, Buffalo, New York 14215*

[2]*Massachusetts General Hospital, Boston, Massachusetts 02114*

We have recently reported the isolation of a cancer-associated serum galactosyltransferase activity which is separated from the normal enzyme by polyacrylamide electrophoresis.[1,2,3] In addition, a small molecular weight glycopeptide (~3600 Daltons), which is a preferred acceptor for the cancer-associated isoenzyme, has been detected in sera of patients with extensive or metastatic diseases.[4] This glycopeptide has a cytotoxic effect on polyoma transformed tissue culture cells and on solid tumors grown in hamsters but has no effect on the non-transformed cell.[5] Our current hypothesis is that both the cancer-associated isoenzyme (GT-II) and glycopeptide acceptor (CAGA) are from the tumor cell, probably shed or secreted. In this article we will review the data which support their association with cancer, indicate their possible clinical value, and discuss the theoretical implications for the biology of tumor development and growth.

The study of human serum galactosyltransferase isoenzymes arose from work on changes in cell surface glycosyltransferases and glycoproteins occurring in association with mitosis, early growth and malignant transformation. Our approach is based on the reports of many investigators which suggest that malignant transformation is closely associated with fundamental, if not prerequisite, changes in the cell surface,[6,7,8] changes similar to those described for embryonic or fetal cells.[9-12] In many of these studies the major alterations were shown to be in the polysaccharide portion of plasma membrane glycolipids and glycoproteins. Similar changes were also found in our earlier work on cell surface changes associated with differentiation of rat intestinal epithelium.[13-16]

One of our findings was that the enzymes responsible for the synthesis of polysaccharide chains could be detected on the cell surface of mitotically active intestinal crypt cells. These enzymes which have the general reaction

$$\text{nucleotide-sugar} + \text{acceptor} \xrightarrow{\text{cation}} \text{sugar-acceptor} + \text{nucleotide}$$

were previously considered marker enzymes for Golgi and smooth endoplasmic reticulum. Thus, when Roth, McGuire and Roseman demonstrated glycosyltransferases on the cell surface of embryonic chicken neural retinal cells growing in tissue culture,[17] Keenan and Moore challenged the concept that glycosyltransferases exist on the surface membrane of mammalian cells and insisted that they were present only on Golgi membranes.[18] Most recent publications support glycosyltransferases being on the cell surface[19,20] of some cells but their roles, if any, have not been defined.

When we observed that glycosyltransferase activities were present on the surface of the mitotically active intestinal crypt cell, we reasoned that their appearance on the plasma membrane could be the result of the cell's sudden need, during mitosis, for new cell membrane. Since other investigators had previously suggested that the cell membrane was derived from the Golgi apparatus we concluded that the "early cell surface membrane is, in fact, externalized Golgi membrane."[15] Patt and Grimes have since suggested a similar mechanism for the appearance of glycosyltransferases on the cell surface membranes of tissue culture cells that is related to cell cycle and viral transformation.[21] Although there continues to be less controversy on the existence of plasma membrane glycosyltransferases, there is greater doubt as to whether any correlation exists between the appearance of cell surface glycosyltransferases and malignant transformation. Nevertheless, it was this hypothesis that led us to look for cancer-associated glycosyltransferase enzymes in patients' serum.

Serum galactosyltransferase isoenzymes. During an investigation on the relationship between cell surface galactosyltransferase and conconavalin A agglutination of cells, we solubilized and purified a galactosyltransferase from rabbit erythrocyte membranes.[22] That purification involved a polyacrylamide electrophoretic step where it was demonstrated that the enzyme could be eluted from the gel. This finding prompted us to re-investigate the possible association of cancer with galactosyltransferase activity. Other investigators had not shown any correlation of total serum galactosyltransferase with cancer, but we reasoned that tumor cells might be shedding an altered or intrinsically different galactosyltransferase into the serum and that it could be detected with separation on polyacrylamide electrophoresis. Although we have not rigidly proven that any serum galactosyltransferase activity originates from tumor cells, we have isolated an isoenzyme of serum galactosyltransferase activity which appears to correlate with cancer.

This isoenzyme is referred to as isoenzyme II or GT-II and was detected as a slower moving peak of galactosyltransferase activity on polyacrylamide gel electrophoresis. This peak of galactosyltransferase activity separated well from the major, more anodally directed peak of activity, isoenzyme I or GT-I, found in all subjected tested (Figure 1).

In our initial studies GT-II was detected in approximately 75% of 80 patients tested with various types of cancer (Figure 2). Twenty-two normal subjects and 52 patients with non-malignant disorders did not demonstrate GT-II except for 3 of 15 patients with severe alcoholic hepatitis, one elderly patient with diverticulitis and 18 out of 20 (latest data) patients with relatively active celiac disease. One of these patients with celiac disease, who demonstrated GT-II at the time of initial diagnosis (subtotal mucosal atrophy), did not show GT-II activity after one year of treatment and a complete return to a normal mucosa.

Later studies[3] when totaled with our earlier studies showed that of 232 patients with cancer, 71% demonstrated GT-II. The later studies concentrated on colon cancers and showed a correlation between the extent of disease at operation and the level of GT-II activity (Figure 3). Of nine patients with colon cancer that did not penetrate the serosa but was present in the muscularis (Dukes' B), seven showed GT-II in their serum which was not present after colectomy. We followed 20 patients who initially did demonstrate GT-II activity and noted a positive correlation between progression of disease and an increase in GT-II activity (Figure 4).

Animal tumor models demonstrating GT-II. As we carried out the clinical studies, experiments with animal tumor models were initiated. One model uses polyoma transformed BHK cells to form solid tumors in the backs of hamsters.[23] Growth of the tumors correlated with the appearance in serum of an electrophoretically distinct peak of galactosyltransferase activity which ran slower than the bulk of activity from non-tumor control hamsters, very similar to the pattern seen in patients. This slow moving peak ($GT-II_H$) was detectable before solid tumors could be grossly observed, and the amount of activity in this peak increased with growth of the tumor (Figure 5). $GT-II_H$ was not detectable in control animals and separated from a faster migrating major area of serum galactosyltransferase activity ($GT-I_H$) found in sera of both control and tumor-bearing hamsters. These two activites were shown to maintain their respective mobilities on re-electrophoresis. Solubilized enzyme derived from excised tumors demonstrated an electrophoretic mobility identical to that for serum $GT-II_H$ found in tumor-bearing animals. In contrast, enzyme activity solubilized from livers of both control and tumor-bearing hamsters showed a mobility similar to that of the fast moving $GT-I_H$ (i.e. the normal peak).

Another animal tumor model, in which human mammary adenocarcinoma cells were grown in athymic nude mice, also demonstrated the emergence of a tumor-associated electrophoretically distinct peak of galactosyltransferase activity in the serum of tumor-bearing mice. This galactosyltransferase isoenzyme was detected in the tumor itself suggesting again that tumor was the source for the cancer-associated isoenzyme.

Purification of human cancer-associated GT-II. The term isoenzyme implies a fundamental structural alteration in the protein; to demonstrate this, purification was necessary. Attempts to extract sufficient and stable activity from tumors was unsuccessful. Malignant effusions proved to be the best source for purification of both the normal and cancer-associated isoenzymes.[24] It is interesting to note that GT-II was only present in those effusions which were cytology positive (Table 1), again suggesting that the tumor cell itself is the source for the isoenzyme. The two isoenzymes could be separated by DEAE chromatography and by a series of affinity chromatographic steps; each enzyme was purified approximately 5000 fold. The cancer-associated isoenzyme differed from the normal one by carbohydrate, amino acid composition and peptide mapping. Thus, it appears that GT-II is an isoenzyme of GT-I. Of particular interest was the demonstration that the cancer-associated GT-II had much less affinity for the substrate SGF-fetuin than did the normal GT-I. In contrast, GT-II showed a greater affinity for a cancer-associated glycopeptide acceptor which we had detected in patients with extensive disseminated cancers.[5]

A cancer-associated glycopeptide acceptor activity. The discovery of this preferred acceptor for GT-II was initially detected as an inhibitory activity in the usual galactosyltransferase assay using SGF-fetuin as acceptor where the product is detected as acid precipitable radioactivity. We had observed a puzzling loss of GT-II activity in the tumor-bearing hamsters when the tumors were extensive (over 35 days after inoculation). Mixing experiments suggested an "inhibitor" of galactosyltransferase activity in the sera of these animals with huge tumor loads. A similar phenomenon was seen in patients with extensive cancer. It was subsequently demonstrated that this inhibition was partly due to a glycopeptide which was a competitive acceptor for GT-II. This has not accounted for the loss of GT-II after electrophoresis, which is under current investigation. Preliminary data suggested that sera containing this glycopeptide affected transformed cells, and we purified this glycopeptide from malignant cytology-positive effusions.[4]

The glycopeptide is 70% carbohydrate by weight and serves as a preferred acceptor for GT-II as compared to GT-I. Its molecular weight is approximately 3600 and appears to be a carbohydrate-enriched remnant of an asparagine-linked glycoprotein. The tissue source is unknown and although we favor a tumor-cell origin, another cell source such as a plasma cell is also possible. In the animal tumor model, its level increases as the tumor size becomes overwhelming and as GT-II levels appear to decrease (Figure 5).

This glycopeptide which we have termed CAGA for cancer-associated galactosyltransferase activity is unusual in that it appears to be cytotoxic for virally transformed cells in tissue culture and does not affect the normal counterpart (at concentrations up to 10 μg per dish).[5] In addition, when purified CAGA was given to tumor-bearing animals, a reduction in tumor mass was observed.

Biological implications. The biological function and clinical utility of GT-II and CAGA need further exploration. They may have no biological function but GT-II may be useful in the diagnosis or confirmation of cancer, and CAGA's preferential cytotoxicity for virally transformed cell lines and human tumor cells in tissue culture indicate the need for further pharmacological testing. The mechanism of the cytotoxicity needs to be defined.

The relationship of CAGA to GT-II may be incidental but it is intriguing to note that CAGA has at least one order greater affinity for GT-II than for GT-I. If both are tumor products, they may interact to enhance metastases and "fool" the host defense mechanisms. We have recently postulated such an hypothesis for GT-II.[25]

Our speculations on GT-II centered on a role for GT-II in terms of cell adhesion forces. These speculations were based on some early concepts proposed by Roseman on the role of glycosyltransferases in cell adhesion for which there is, as yet, no evidence.[8] Our working hypothesis is that GT-II, a glycoprotein enzyme, serves as a recognition factor in the growth and metastases of tumor cells. It is tumor-self which can adhere to a normal cell and be recognized by the tumor cell as a receptive site. We speculate a divalent character such as occurs in lectin agglutination which permits the tumor cell to bind to its own GT-II which in turn has been able to bind to a normal cell (Figure 6).

Acknowledgements

This work was supported by grants from the USPHS (CA 14294) and American Cancer Society (BC-93 and PDT-88). During part of this study, Dr. Weiser was a research investigator for the Howard Hughes Medical Institute.

REFERENCES

1. Podolsky, D.K., Weiser, M.M.: Biochem. Biophys. Res. Comm. 65:545, 1975.
2. Weiser, M.M., Podolsky, D.K., Isselbacher, K.J.: Proc. Natl. Acad. Sci. (USA) 73:1319, 1976.
3. Podolsky, D.K., Weiser, M.M., Isselbacher, K.J., Cohen, A.M.: NEJM 299:703, 1978.
4. Podolsky, D.K., Weiser, M.M.: Biochem.J. (in press)
5. Podolsky, D.K., Weiser, M.M., Isselbacher, K.J.: Proc. Natl. Acad. Sci. (USA), 75:4426, 1978.
6. Cook, G.M.W.: Biol. Rev. 43:363, 1968.
7. Fox, T.O., Shepard, J.R., Burger, M.M.: Proc. Natl. Acad. Sci. (USA), 68:244, 1971
8. Roseman, S.: Chem. Phys. Lipids, 5:270, 1970.
9. Coggin, J.H., Jr., Ambrose, K.R., Anderson, N.G.: J. Immunol. 105:524, 1970.
10. Maugh II, T.H.: Science 184:147, 1974.
11. Pearson, G., Freedman, G.: Cancer Res., 28:1665, 1968.
12. Duff, R., Rapp, F.J.: Immunol. 105:521, 1970.
13. Weiser, M.M.: Science 177:525, 1972.
14. Weiser, M.M.: J. Biol. Chem. 248:2536, 1973.
15. Weiser, M.M.: J. Biol. Chem. 248:2542, 1973.
16. Podolsky, D.K., Weiser, M.M.: J. Cell Biol. 58:497, 1973.
17. Roth, S., McGuire, E.J., Roseman, S.: J. Cell Biol. 51:536, 1971.
18. Keenan, T.W., Morré, D.J.: FEBS Letters 55:8, 1975.
19. Shur, B., Roth, S.: Biochim. Biophys. Acta 415:473, 1975.
20. Weiser, M.M., Neumeier, M.M., Quaroni, A., Kirsch, K.: J. Cell Biol. 77:722, 1978.
21. Patt, L.M., Grimes, W.J., J. Biol. Chem. 249:4157, 1974.
22. Podolsky, D.K., Weiser, M.M.: Biochem. J. 146:213, 1975.
23. Podolsky, D.K., Weiser, M.M., Westwood, J.C., Gammon, M.: J. Biol. Chem. 252:1807, 1977.
24. Podolsky, D.K., Weiser, M.M.: J. Biol. Chem. (in press).
25. Weiser, M.M., Podolsky, D.K. in Cell Surface Carbohydrate Chemistry. Edited by R.E. Harmon. (Academic Press, Inc., New York, 1978) pp. 67-82.

TABLE 1. CORRELATION OF THE PRESENCE OF GT-II WITH CYTOLOGY POSITIVE EFFUSIONS.

Type of effusion	Effusion GT-II activity
	pmol/25 µl/60 min
Cancer:	
cytology positive (28)	3.7 ± 0.5
cytology positive (4)	0
cytology negative (12)	0
Non-cancer disease (8)	0

Figure 1: A representative polyacrylamide electrophoretic pattern of serum galactosyltransferase activity. Arrow points to slower moving peak found in serum of a patient with proven carcinoma of the colon.

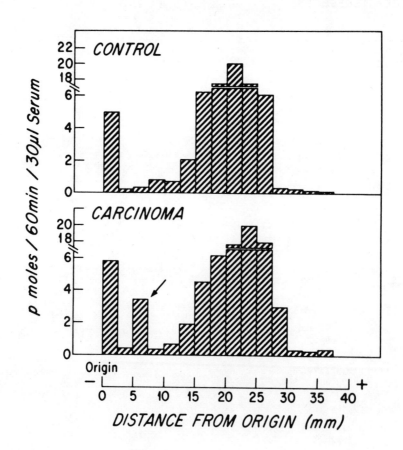

Figure 2: Comparison of human serum galactosyltransferase isoenzyme I and II levels among normal controls, patients with carcinoma and patients with diseases other than cancer.

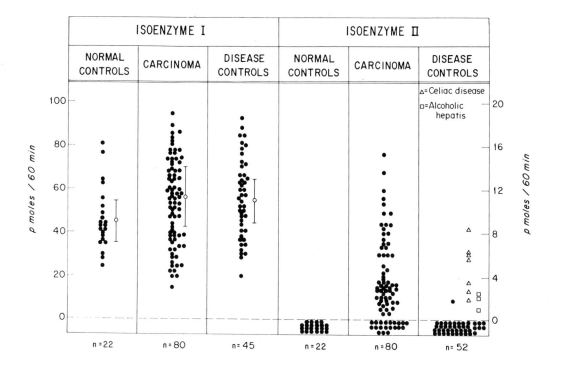

Figure 3: Galactosyltransferase II levels in patients with colorectal carcinoma; relationship to size of tumor burden.

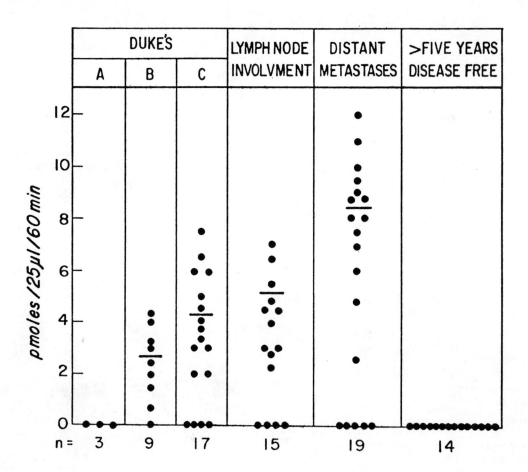

Figure 4: Serial evaluation of galactosyltransferase II in patients with colorectal cancers.

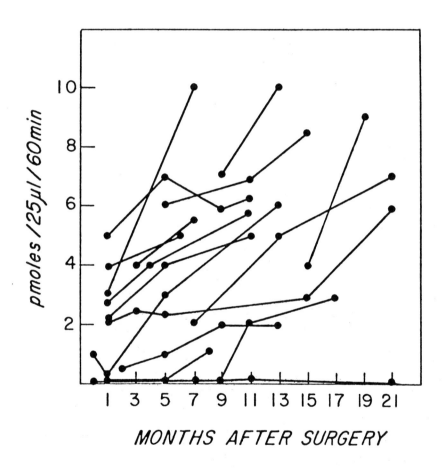

Figure 5: Hamster serum galactosyltransferase II levels after inoculating virally transformed cells capable of forming tumors. The first part of the graph demonstrates that GT-II increases as the cell inoculation grows. (O) and (●) represent two different types of polyoma transformed BHK cells. The right half of the graph shows the loss of GT-II activity in one group of hamsters as their tumor mass becomes overwhelming. Concomitantly there is detected an inhibitor of galactosyltransferase activity (▲).

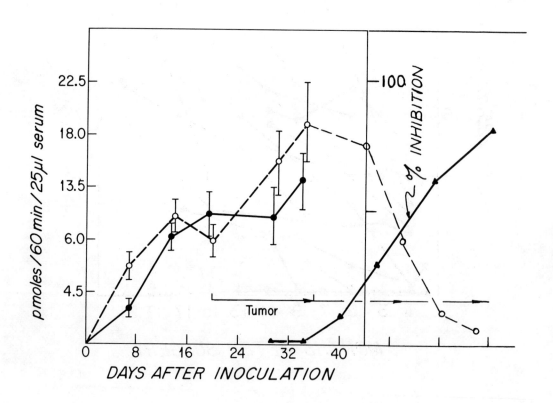

Figure 6:

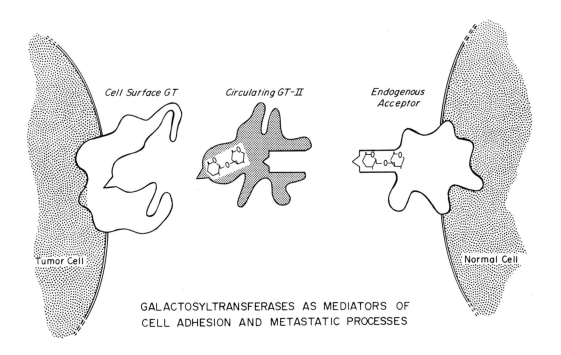

GALACTOSYLTRANSFERASES AS MEDIATORS OF CELL ADHESION AND METASTATIC PROCESSES

IMMUNOCYTOLOGICAL DEMONSTRATION OF GROWTH HORMONE IN MAMMARY CARCINOMA CELLS

Bimal C. Ghosh, M.D.[2] and Luna Ghosh, M.D.[1]
[1]Associate Professor, Department of Pathology
[2]Associate Chairman, Division of Surgical Oncology, [1]Associate Professor of Surgery, Department of Surgery, University of Illinois, Abraham Lincoln School of Medicine and Cook County Hospital, Chicago, Illinois

INTRODUCTION

Carcinoma of the breast is a major problem in the United States. The incidence is rising and it is estimated that 90,000 new cancers of the breast will develop in the year of 1978. On the other hand, one out of fifteen women will develop cancer of the breast; every 15 minutes, three women will develop this disease and one will die.

It has been known for years that the breast carcinoma has a relation to the endocrine glands (1, 4, 5). The hormone sensitivity of the breast tumor has been studied in both male and female subjects. The hormone receptor site in tumor has been investigated by autoradiographic technique; the exact cytological demonstration of hormones in the mammary carcinoma cells has not been shown. The objective of this study is to localize hormones within the mammary carcinoma cells by using the unlabeled antibody enzyme technique with horseradish and antihorseradish peroxidase complex (2). The growth hormone was selected to investigate whether any other hormone of the pituitary which has not been investigated has a role in breast carcinoma. In 1966, Paul Nakane first developed a technique by using enzyme-labeled antibodies for localization of hormones. He employed antibodies which were conjugated to peroxidase by a bifunctional reagent 4-4 difluro-3-3 dinitro-diphenyl-sulfone. After application of conjugated antibodies to the tissue, the peroxidase was localized histochemically with hydrogen peroxidase and 3-3 diaminobenzidine. Then, the tissue was osmicated to enhance the reaction product. In 1973, Nakane improved his technique when he was experimenting with his conjugated antibodies. Two different groups of investigators modified the immunoperoxidase technique. One group, Mason, Phifer, and Spicer (3) from South Carolina, in 1969, called it immunoglobulin enzyme bridge technique. The second group, Sternberger, et al. (6), from John Hopkins University, School of Medicine, called it unlabeled antibody enzyme technique. Both of these methods needed the components:
1) Specific antibodies to the tissue antigens made in a species such as rabbits.
2) Goat or sheep antiserum against immunoglobulin of the species in which antiserum was produced such as sheep anti-rabbit gammaglobulin.

3) Anti-peroxidase antibodies made in the same species as component one, such as rabbit.
4) Horseradish peroxidase.

Peroxidase is localized histochemically with hydrogen peroxidase, DAB, and osmium tetroxide. Initially, we used this technique to localize the growth hormone and C_3H mouse mammary adenocarcinoma cells. After we standardized the technique, we utilized this procedure in human mammary carcinoma cells.

MATERIALS AND METHODS

Tissues obtained from male and female patients with infiltrating duct carcinoma were fixed in 2% calcium acetate formalin and subsequently dehydrated through graded alcohol; sections were cut and mounted on glass slides. Normal female breast tissues were similarly treated as control. For growth hormone localization, the sections were immersed in phosphate buffer saline at pH 7.2 for half an hour before treatment with antisera. Two drops of rabbit anti-growth hormone in different dilutions were added to each section and they were kept in a moist chamber at 37°C for one hour. Sheep anti-rabbit gammaglobulin was added to each section and the sections were again kept in 37°C in a moist chamber for one hour and washed with phosphate buffer saline three times. Rabbit peroxidase anti-peroxidase complex was added to this section for one hour at the same temperature. Sections were again washed with phosphate buffer and incubated in the substrate solution for five minutes at room temperature. The substrate solution was composed of 30 mg per 100 ml of 3-3 diaminobenzidine-tetrahydrochloride in Tris buffer. The buffer solution was made with 60 mg per 100 ml of Tris hydroxymethylaminometham in phosphate buffer adjusted to pH 7.6. A few drops of 3% hydrogen peroxidase were added just before the experiment. After the initial reaction, the section was washed in phosphate buffer saline, then with distilled water, then osmicated for ten minutes in 2% osmium tetroxide. It was washed again in distilled water, dehydrated, and mounted in parmount. The cytoplasm of the mammary carcinoma cells was stained when sections were treated with anti-growth hormone serum. The positive reaction showed a blackish-brown coloration which was observed throughout the cells. In male carcinoma cells, the reactions were very intense; female carcinoma cells showed similar staining reaction with a definite affinity for cytoplasmic membrane. The normal female breast tissue showed a very mild reaction; red blood cells showed endogenous peroxidase activity. Control study was done by substituting normal sheep serum for sheep anti-rabbit serum and some sections were incubated in diaminobenzidine peroxidase only, all those failed to show any positive reaction.

COMMENTS

In clinical practice, often it is observed that the influence of hormones on the malignant tumors is present although the exact mechanism and the site of interference are unknown. Although with the use of radioautography, the receptor sites of several hormones have been known, it is not possible to know which cellular component of cancer is influenced by the hormones. The present technique might selectively localize the site of activity of the specific hormone in the malignant tumor cells. The presence of these hormones or hormone-like substances in the different sites of the malignant tumor cells might have some prognostic influence on these particular types of tumors. The question whether these intracellular localizations are truly receptor sites or whether some of the malignant cells independently secrete this hormone or hormone-like substance still remains unanswered.

REFERENCES

1. Block, G.E., Jensen, E.V., Polley, T.Z., Jr.: The prediction of hormonal dependency of mammary cancer. Ann. Surg. 182: 342, 1975.

2. Ghosh, L., Ghosh, B.C., Das Gupta, T.K.: Immunocytological localization of estrogen in human mammary cells by horseradish-anti-horseradish peroxidase complex. J. Surg. Oncol. 10:221, 1978.

3. Mason, E.E., Phifer, R.F., Spicer, S.S., Swallow, R.A., Dreskin, R.S.: An immunoglobulin-enzyme bridge method for localizing tissue antigens. J. Histochem. Cytochem. 17:563, 1969.

4. McGuire, W.L.: Current status of estrogen receptors in human breast cancer. Cancer 36:638, 1975.

5. Mobbs, B.G.: Estradiol uptake by induced rat mammary tumors and its implications for the treatment of breast cancer. Nat. Cancer Inst. Monogr. 34:33, 1971.

6. Sternberger, L.A., Hardy, P.H., Cuculis, J.J., Meyer, H.G.: The unlabeled antibody enzyme method of immunohistochemistry. Preparation and properties of soluble antigen-antibody complex (horseradish peroxidase-antihorseradish peroxidase) and its use in identification of spirochetes. J. Histochem. Cytochem. 18:315, 1970.

PROSTAGLANDIN PRODUCTION BY MURINE TUMORS

Scott W. Burchiel, Ph.D.[1,2] Mark Rubin, B.S.[2,3] Janis Giorgi, Ph.D.[2] Glenn T. Peake, M.D.[4] Noel L. Warner, Ph.D.[2]

[1]*University of New Mexico, College of Pharmacy,* [2]*University of New Mexico School of Medicine, Immunobiology Labs,* [3]*Columbia University School of Medicine, New York,* [4]*University of New Mexico School of Medicine, Department of Medicine, University of New Mexico Medical Center, Albuquerque, New Mexico 87131*

INTRODUCTION

Previous studies have shown that a variety of human and animal tumors have the ability to synthesize and secrete E type prostaglandins, (PGE).[1] Since many tumors have also been shown to induce a state of non-specific immunosuppression, it has been considered by several authors[2,3] that this immunosuppression may, in turn, be due to the action of PGE which is itself a recognized immunosuppressive agent.[4,5,6,7]

In previous studies on the induction of cytotoxic T cells to tumor specific antigens, it was observed that many tumors would stimulate only at low stimulator to responder cell ratios[8,9] suggesting that at high concentrations of tumor cells, immunosuppression might be operating. In further studies on this series of tumors, direct suppression by the tumor cells was demonstrated by the addition of these cells to mixed lymphocyte cultures.[10] This present report analyzes the PGE synthesis by these tumors and its possible relation to immunosuppression. These cell lines that were found to secrete significant amounts of PGE were also tested for their ability to suppress an *in vitro* mitogen-induced lymphocyte proliferative response in the presence or absence of a PGE synthesis inhibitor. The results of this study showed that although certain cell lines secreted PGE and were immunosuppressive, that PGE was not likely to be the principal active agent in suppressing the mitogen response. Thus, our results suggest that other unidentified factors are responsible for the suppressive effects of these tumors on *in vitro* T and B cell responses.

MATERIALS AND METHODS

<u>Tumor cell culture</u>: All tumor cells used in this study were from established cell lines grown in suspension culture in Dulbecco's modified Eagles medium (DME) with 10% fetal calf serum. Certain B lymphoma and macrophage lines also required 2 mercapto ethanol and/or nonessential amino acids for growth. All cells used for experiments were taken during log stage of growth and were greater than 95% viable.

<u>Assay of PGE production</u>: The production of PGE by cultured cell lines was assessed by taking tumor cells during the log stage of

growth, washing them one time with culture media, followed by resuspension in fresh culture media at a density of 3×10^5/ml, and incubation for 24 hours in a 10% CO_2 incubator at 37^0 C. Following this incubation, culture supernatants were harvested and were extracted 5 times with freshly distilled ethyl acetate. The organic fractions were then pooled and were evaporated under a stream of N_2 at 43^0 C. These extracts were then stored under argon at -20^0 C until ready for assay. The PGE content of the ethyl acetate extract was then evaluated by resuspending the extracts in phosphate buffered saline pH 7.0 and then performing a radioimmunoassay as previously described.[11,21] The antiserum used showed complete cross reactivity between PGE_1 and PGE_2, and approximately 70% cross reactivity with PGA.[21]

In vitro proliferative response: 2×10^5 spleen cells taken from 8-12 week old BALB/c ByJ mice were cultured in microtiter plates in DME + 10% FCS for 72 hours with an optimal mitogenic dosage of Con A (3 µg/ml) or LPS (50 µg/ml), in the presence or absence of x-irradiated (5000 rad) syngeneic tumor cells. 1 µCi of ^{3}H-thymidine was added during the last 18 hours of cell culture to determine the level of DNA synthesis. The total amount of ^{3}H-thymidine incorporation was used as an index of cellular proliferation and was compared with cells not exposed to mitogenic agents and/or tumor cells. In certain experiments, indomethacin was present in the culture at a final concentration of 10^{-5} M.

RESULTS AND DISCUSSION

When a variety of cultured murine tumors were examined for prostaglandin production, considerable differences were observed between tumors. These assays have been repeated several times, and the tumors are grouped into three classes according to their ability to synthesize and secrete PGE (Table 1). The mammary carcinoma, EMT-6, secreted very high concentrations of PGE, and the fibrosarcomas also made significant amounts of PGE. The majority of the lymphoid and myeloid tumors tested however secreted little, if any, PGE. This finding of a high level of PGE production by a mammary carcinoma and moderate levels from fibrosarcomas is in general accord with the finding of Tan, *et al*. that rat mammary tumors secrete PGE_2[12] and of other investigators who have also shown that mouse fibrosarcomas secrete PGE_2.[13,14] Thus, our findings concerning the PGE production by murine tumors are in general agreement with previously published results.

Although only the mammary carcinoma and the fibrosarcoma cell lines examined were capable of synthesizing and secreting detectable levels of PGE, some cell lines could be induced to secrete PGE when exposed to LPS and/or zymosan (Table II). Those cell lines that were found to have inducible PGE production all belonged to the macrophage-monocytic cell lineage. It is notable that peritoneal macrophages taken from rats,[15] human blood monocytes, and mouse peritoneal macrophages[16,17] have been shown to have inducible PGE production In addition, two of the cell lines that were tested, WEHI-3 and J774, have been previously observed to have LPS-inducible PGE production.[16] Thus, these results not only serve to confirm the findings of inducible PGE production, but also point out the need to examine PGE production in "activated" as well as "resting" populations of neoplastic cells. In some experiments low, but possibly significant, levels of PGE were detected from the non-stimulated macrophage cell lines. A further finding of the present study is that LPS and zymosan show a a differential ability to induce PGE production in macrophage-monocytic cell lines. Whereas, LPS and zymosan were both effective in inducing PGE production in WEHI-3, PU-5, and J774; only zymosan acted as an inducing agent with P388D1.

The differential ability of LPS and zymosan to induce PGE production in these cell lines may reflect differences in cell line differentiation states, and is the subject of further investigation.

Two of the cell lines that were found to secrete moderate to high concentrations of PGE were tested for their effects on the *in vitro* mitogen-induced spleen cell proliferative response. The particular assay assessed the ability of EMT-6 and WEHI-164 to affect the Con A and LPS proliferative response of syngeneic s spleen cells as measured by ^{3}H-thymidine incorporation. Representative experiments show that EMT-6 inhibits the Con A response of spleen cells in a dose-dependent manner (Figure 1). However, the ability of EMT-6 to inhibit Con A-induced spleen cell proliferation was not altered by the presence of 10^{-5} M indomethacin, an effective prostaglandin synthesis inhibitor. Because we found that at the concentration used indomethacin completely inhibited PGE production by EMT-6 (data not shown), we conclude that there is a lack of correlation between the ability of EMT-6 to produce PGE and its ability to inhibit *in vitro* reactivity of cultured spleen cells. This conclusion was further supported by our findings with WEHI-164 using both a Con A (Figure 2) and LPS (Figure 3) induced system. Again, our results show that indomethacin did not prevent suppression of mitogen-induced spleen cell proliferation produced by WEHI-164. Similar results have also been observed with the macrophage tumors P388.D1 and J774 which, although secreting only marginal levels of PGE, were very immunosuppressive, either with or without indomethacin being present (Table 3).

The results of the present study indicate that PGE production by tumors is not likely to account for the mechanism by which these tumor cells produce suppression of the *in vitro* mitogen-induced proliferative response. This conclusion is in general agreement with the findings of Harvey, *et al.* who showed that anergic cancer patients did not have elevated levels of serum PGE.[18] However, it is clear that prostaglandin production can contribute to the overall pathology observed with tumors that make PGE.[15] For example, certain types of hypercalcemia have been associated with prostaglandin secreting tumors.[20] Whether indirect effects of prostaglandin production, such as hypercalcemia, contribute to effects seen *in vivo* on immunologic reactivity is unclear. It is notable that Plescia, *et al.* have found that indomethacin retards tumor growth and restores a degree of immunologic responses in mice implanted with MC-16 fibrosarcoma.[2] This effect was also seen *in vitro*.[3] In an occasional experiment with EMT-6 and WEHI-164, we have observed that some slight inhibition of the suppressive effect occurred in the presence of indomethacin, but only at very low tumor cell numbers. In addition, Goodwin, *et al.* have found that patients with Hodgkin's disease are anergic, and have restoration of *in vitro* reactivity when cultures are treated with indomethacin.[21] We, therefore, conclude that although prostaglandins may be involved in suppression of the immune response by some tumors, it is not a common denominator for suppression produced by all tumors, nor does it account for all (if any) of the suppression mediated by many PGE secreting tumors. The relative contribution of prostaglandin production to the decreased immunologic reactivity observed in association with many tumors must therefore be further assessed.

REFERENCES

1. Karim, S.M.M., Rao, B.: Prostaglandins and tumors, IN, Prostaglandins: Physiological, Pharmacological and Pathological Aspects. Ed. Karim, S. M.M., p. 293, University Park Press, Baltimore, 1976.
2. Plescia, O.J., Smith, A.H., Grinwich, K.: Subversion of the immune system by tumor cells and role of prostaglandins. Proc. Nat. Acad. Sci. 72:1848, 1975.
3. Grinwich, K.D., Plescia, O.J.: Tumor mediate immunosuppression: prevention by inhibitors of prostaglandin synthesis. Prostaglandins 14:1175, 1977.
4. DeRubertis, F., Zenser, T.V., Adler, W.H., Hudson, T.: Role of cyclic adenosine 3':5' - monophosphate in lymphocyte mitogenesis. J. Immunol. 113:51, 1974.
5. Gordon, D., Bray, M.A., Morley, J.: Control of lymphokine secretion by prostaglandins. Nature 262:401, 1976.
6. Bourne, H.R., Lichtenstein, L.., Melmon, K.L.: Pharmacologic control of allergic histamine release in vitro: Evidence

for an inhibitory role of 3' : 5' - adenosine monophosphate in human lymphocytes. J. Immunol. 108:695, 1972.

7. Webb, D.R., Nowowiejski, I.: The role of prostaglandins in the control of the primary 19S immune response to SRBC. Cell Immunol. 33:1, 1977.
8. Burton, R.C., Thompson, J., Warner, N.L.: In vitro induction of tumor specific immunity. I. Development of optimal conditions for induction and assay of cytotoxic lymphocytes. J. Immunol. Meth. 8:133, 1975.
9. Burton, R.C., Warner, N.L.: In vitro induction of tumor specific immunity V. Detection of common antigenic determinants of murine fibrosarcomas. Brit. J. Cancer 37:159, 1978.
10. Warner, N.L., Giorgi, J.V., Daley, M.J.: In vitro induction of tumor specific immunity, IN Proceedings of U.S.-Japan Cancer Immunology Conference, Miami, Hawaii, 1979.
11. Rigler, G.L., Peake, G.T., Ratner, A.A.: Effects of follicle-stimulating hormone and luteinizing hormone on ovarian cyclic AMP and prostaglandin E in vivo in rats treated with indomethacin. J. Endocrinol. 70:285, 1976.
12. Tan, W.C., Privett, O.S., Goldyne, M.E.: Studies of prostaglandins in rat mammary tumors induced by 7, 12-dimethylbenz(a)antracene. Cancer Res. 34:3229, 1974.
13. Tashjian, A.H., Voelkel, E.F., Goldhaber, P., Levine, L.: Successful treatment of hypercalcemia by indomethacin in mice bearing a prostaglandin producing fibrosarcoma. Prost.3:515, 1973.
14. Levine, L.: Prostaglandin production by mouse fibrosarcoma cells in culture: Inhibition by indomethacin and aspirin. Biochem. Biophys. Res. Comm. 47:888, 1972.
15. Gemsa, D., Seitz, M., Kramer, W., Till, G., Resch, K.: The effects of phagocytosis, dextran sulfate, and cell damage on PGE_1 sensitivity and PGE_1 production of macrophages. J. Immunol. 120:1187, 1978.
16. Kurland, J.I., Bockman, R.: Prostaglandin E production by human blood monocytes and mouse peritoneal macrophages. J. Exp. Med.147:952, 1978.
17. Humes, J.L., Bonney, R.J., Pelus, L., Dahlgren, M.E., Sadowski, S.J., Kuehl, F.A., Davies, P.: Macrophages synthesize and release prostaglandins in response to inflammatory stimuli. Nature 269:149, 1977.
18. Harvey, H.A., Allegra, J.C., Demers, L.M., Luderer, J.R., Brenner, D.E., Trautlein, J.J., White, D.S., Gillin, M.A., Lipton, A.: Immunosuppression and cancer: Role of prostaglandins. Cancer 39:2362, 1977.
19. Haffe, B.M.: Prostaglandins and cancer: An update. Prost. 6:453, 1974.
20. Seyberth, H.Y., Segre, G., Morgan, J.L., Sweetman, B.J., Potts, J.T., Oates, J.A.: Prostaglandins as mediators of hypercalcemia associated with certain types of cancer. NEJM 293:1278 1975.

21. Goodwin, J.S., Messner, R.P., Bankhurst, A.D., Peake, G.T., Saidi, J.H., and Williams, R.C.: Prostaglandin producing suppressor cells in Hodgkin's disease. <u>NEJM</u> 297:963, 1977.

Supported by U.S.P.H. Grants NIH CA 22268, HD0594, RR0997, American Cancer Society grant IM-175, and the KROC Foundation.

Acknowledgements:
The expert technical assistance of Ms. Jeane Frey, Ms. Kirsten Hanson, and Mr. Keith Rasmussen is gratefully acknowledged.

Figure 1. Dose-dependent inhibition of spleen cell ^{3}H-thymidine incorporation produced by EMT-6 cells in the presence or absence of 10^{-5}M indomethacin.

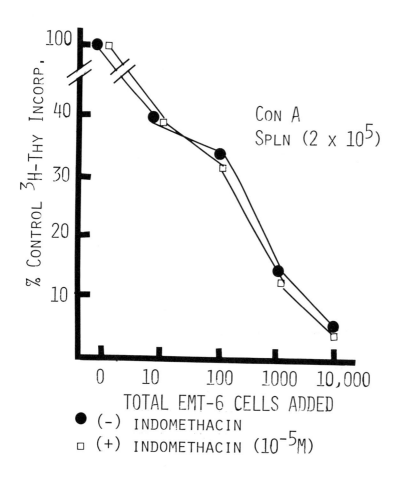

Figure 2. Inhibition of Con A-induced spleen cell proliferation produced by WEHI-164 cells in the absence or presence of indomethacin.

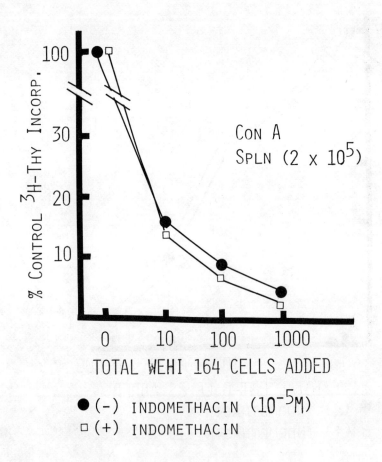

Figure 3. As in Figure 2, except that spleen cells were stimulated with an optimally mitogenic dose of LPS.

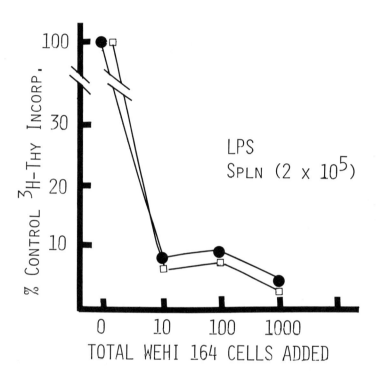

TABLE I. PGE PRODUCTION BY CULTURED MURINE TUMOR CELL LINES[a]

Cell Line	Cell Type	PGE Secretion Range ng/ml/10^6 cells
EMT-6	Mammary Carcinoma	> 100
WEHI-164	Fibrosarcoma	10-100
WEHI-11	Fibrosarcoma	10-100
WEHI-3, PU-5. 1R		
P388D1, J774	Macrophage-Monocytic	
WEHI-274		
MPC-11; MPC-11. NP; MPC 11L	Plasma Cell	
HPC-108		< 5
ABE-8 WEHI-279	B Lymphoma	
S49, WEHI-22, WEHI-7	T Lymphoma	
E1-4, YAC-1		

[a] results from 3 experiments

TABLE II.   INDUCIBILITY OF PGE PRODUCTION

Cell Type (Treatment)	ng/ml/10^6 Cells
WEHI-3	0.1[a]
" + LPS[b]	0.4
" + zymosan	0.8
PU-5.1R	2.0
" + LPS	14.0
" + zymosan	5.9
P388 DI	1.8
" + LPS	1.2
" + zymosan	3.9
J774	0.2
" + LPS	0.6
" + zymosan	0.4

[a]values are ± 15%; [b]LPS = 50 µg/ml; [c]zymosan = 50 µg/ml

Cell lines were cultured for 24 hours in the presence or absence of the inducers prior to harvesting tumor cell supernatants for assay.

TABLE III. SUPPRESSION OF CON A MITOGENESIS BY MACROPHAGE TUMORS

Spleen Cells[a]	Tumor Cells[b]	Indomethacin[c]	Stimulation Index[d]	Per cent Suppression[e]
+	–	–	28.6	–
+	–	+	14.0	–
+	J774	–	0.3	98.9
+	"	+	0.2	98.6
+	P388.D1	–	0.2	99.3
+	"	+	0.1	99.3
+	WEHI-3	–	15.3	46.5
+	"	+	9.6	31.4

[a] All cultures contained 2×10^5 spleen cells and 3 g/ml Con A.

[b] Presence or absence of 10^4 tumor cells (5,000 rad).

[c] Presence or absence of 10^{-5} M indomethacin.

[d] Ratio of CPM from spleen/mitogen/± tumor cells to CPM from control culture.

[e] Percent suppression of the mitogen suppression compared using stimulation index.

EFFECT OF GLUCOCORTICOIDS ON THE PRODUCTION OF ALPHA-FETOPROTEIN AND LACTATE AND MALATE DEHYDROGENASES BY HEPATOMA

T. J. Yoo, Kenneth Kuan, Carl S. Vestling, Chao Y. Kuo

Departments of Medicine and Biochemistry, University of Iowa, and VA Medical Centers, Iowa City, Iowa 52242

INTRODUCTION

It was reported recently that glucocorticoids affect alpha-fetoprotein production by a cell line established from Morris hepatoma 8994.[1] Concentrations of glucocorticoids equal to or higher than 10^{-7} M lead to increased AFP production. These cells also secreted alpha M-fetoprotein into the culture medium, but only after the addition of at least 4×10^{-7} M hydrocortisone or 5×10^{-8} M dexamethasone.[1] Lactate dehydrogenase (LDH) and mitochondrial malate dehydrogenase (MDH) have been isolated in Morris hepatoma.[2-7] This study compares the responses of two hepatoma lines, non-AFP-producing 7777 TC (Tissue Culture derived) and AFP producing 7777 Iowa, to a glucocorticoid treatment with respect to their production of AFP and the enzymes LDH and MDH.

MATERIALS AND METHODS

7777 TC cells and 7777 Iowa were cultured and maintained in our laboratory. The tissue was minced and cultured in plastic culture tissue flasks in an atmosphere of 5% CO_2 and 95% air at 37° C without agitation. M-199 and Lewis medium were used.

AFP was measured by radial immunodiffusion.

On day one, 1×10^6 viable cells in 5 ml of M-199 containing 10% fetal calf serum and antibiotics (pen-strept-fungi-neomycin) were seeded into 60 mm diameter tissue culture dishes. On day two, the supernatants were removed, the dishes washed twice, and 5 ml portions of M-199-10% FCS + PSF-N containing 10^{-4}, 10^{-5} or 10^{-6} M hydrocortisone were added.

At day three, 24-hour plates were processed by pipetting off cells using the medium they were grown in. The cells were counted. At day four, 48-hour plates were processed in the same way as the day three samples. Four day incubation of hepatoma cells was studied as above.

For the seven-day incubation study, cells for the four-day incubation were removed and processed on day five. On day eight cells for the seven-day incubation study were removed and AFP assays for the media were done and the cells were counted.

Enzymes were assayed by the methods described previously. Sonicated cell suspensions were used to yield values for total extractable LDH and MDH.

RESULTS AND DISCUSSION

Tables 1 through 4 show the results of the experiments.

Table 1: When 7777 TC cells were treated with 10^{-4} M, 10^{-5} M and 10^{-6} M hydrocortisone for 24 and 48 hours, there were no changes in AFP production. Hydrocortisone failed to induce AFP production.

Table 2: A similar result is shown for AFP production. However, the mitochondrial plus cytosolic enzymes and LDH were both increased with hydrocortisone treatment during 24 and 48 hour periods.

Table 3: With the four day hydrocortisone treatment, 7777 TC cells showed almost double increments of LDH production and tripled to quadrupled increments of MDH, but these treatments again failed to increase AFP production.

Table 4: The effect of hydrocortisone on 7777 Iowa lines is shown. Again there is about a three-fold increase in both LDH and MDH production. AFP was produced in a quantity of about 491 μg/10^6 cells, but this was decreased to the 20% level when treated with hydrocortisone. Since the mean cell number decreased about 50% at the concentrations of 10^{-4} M and 10^{-5} M of hydrocortisone during the seven day incubation, it appears that hydrocortisone might exert some toxic influence on these cells at these concentrations. There was, however, no increase in AFP production with this treatment. This contrasts with the observed LDH and MDH production.

AFP production by the MH 8994 cell line was doubled by treatment with 4×10^{-7} M hydrocortisone. Dexamethasone treatment was reported to have no effect on AH 66 cells.

De Nechaud, et al., postulated that the glucocorticoid effect in AFP production is the change of differentiation of this tumor cell.[1]

Our study demonstrated that even though the origin of the cells is the same, the response to hydrocortisone is heterogenous-- 7777 Iowa responded and 7777 TC did not. Our 7777 Iowa cell line produced less AFP than was reported with MH 8994. It is possible that only a fraction of the AFP producing population is selected by this treatment, so that less AFP production results. If AFP production is due to the presence of a submetacentric marker composed of no. 7 chromosomes and their short arms,[8] it will be less likely that a short treatment with steroids will induce AFP production.[1] Since both 7777 lines differ in karyotype, a detailed chromosomal structure study might produce some useful information.

There are several possible explanations for increased enzyme activity due to hormone treatment. An increase in tyrosine aminotransferase (TAT) activity after steroid treatment has been associated with increased cytoplasmic levels of functional TAT mRNA.[9] This could either be due to an increase of *de novo* synthesis of these enzymes, or to the elimination of inhibitory factors by the steroid, or to hydrocortisone inhibition of the enzyme breakdown.

In summary, with 10^{-4} M, 10^{-5} M, and 10^{-6} M hydrocortisone treatment of the 7777 TC line for 24 and 48 hours, no AFP production was noted, and LDH MDH increased less than two-fold. After four days incubation, 7777 TC showed nearly a double increase in LDH production and a tripled to quadrupled increase of MDH, but AFP production did not increase. The 7777 Iowa line failed to increase AFP production. However, there was an approximate tripling of levels of both LDH and MDH.

BIBLIOGRAPHY

1. DeNechaud, B., J. E. Becker, and V.R. Potter. 1976. Effect of Glucocorticoids on Fetoprotein Production by an Established Cell Line from Morris Hepatoma 8994. Biochemical and Biophysical Res. Comm. 68:8-15.
2. Brummel, M.C., R.J. Carlotti, L. D. Stegink, J.A. Shepherd and C.S. Vestling. 1975. Amino- and Carboxyl-terminal Analyses of Hepatoma Lactate Dehydrogenase Isozymes. Cancer Res. 35:1278-1281.
3. Carlotti, R.J., G.F. Garnett, W.T. Shieh, A. A. Smucker, C. S. Vestling, and H. P. Morris. 1974. Comparison of Purified Lactate Dehydrogenases from Normal Rat Liver and Morris Hepatomas in Rats and in Culture. Biochimica et Biophys Acta 341:357-365.
4. Kuan, K.N., G.L. Jones and S.C. Vestling. Rapid Preparation of Mitochondrial Malate Dehydrogenase from Rat Liver and Heart. Submitted to Biochemistry, September, 1978.
5. Ryan, L.D., and C.S. Vestling. 1974. Rapid Purification of Lactate Dehydrogenase from Rat Liver and Hepatoma: A New Approach. Arch. Biochem. and Biophys. 160:279-284.
6. Sophianopoulos, A.J., C.S. Vestling. 1962. Malic Dehydrogen-

ase from Rat Liver. Biochem. Prep. 9:102-110.
7. Vestling, C.S. 1975. Lactate Dehydrogenase from Liver, Morris Hepatomas and HTC Cells. Isozymes II Physiological Function, Academic Press, Inc., San Francisco, pp. 87-96.
8. Becker, F.F., S.R. Wolman, R. Asofsky, and S. Sell. 1975. Sequential Analysis of Transplantable Hepatocellular Carcinomas. Cancer Res. 35:3021-3026.
9. Hsieh, W.T. and C.S. Vestling. 1966. Lactate Dehydrogenase from Rat Liver. Biochem. Prep. 11:69-75.

TABLE 1. EFFECT OF TREATMENT WITH HYDROCORTISONE ON AFP PRODUCTION BY 7777 TC CELLS

	CONTROL	10^{-4} M	10^{-5} M	10^{-6} M
I. 24-hour treatment[a]	2.84×10^6 cells[b]	2.11×10^6	2.09×10^6	1.85×10^6
AFP Level[c]	0	0	0	0
II. 48-hour treatment	5.12×10^6 cells[b]	2.15×10^6	2.89×10^6	3.03×10^6
AFP Level	0	0	0	0

a. 24 hours after inoculation of hepatoma cells, the average value of triplicate plates was 1.85×10^6 cell/plate.

b. Both floating and attached cells were counted.

c. AFP was not detected after even 13-fold concentration.

TABLE 2. EFFECTS OF HYDROCORTISONE ON 7777 TC CELLS

24 hours incubation

Concentration	LDH I.U/10^6 cells	MDH I.U./10^6 cells	AFL level µg/10^6 cells	µg/µl
0 (control)	0.278	0.0842	0	0
1×10^{-4} M	0.338	0.1160	0	0
1×10^{-5} M	0.350	0.0838	0	0
1×10^{-6} M	0.336	0.1000	0	0

48 hours incubation

0 (control)	0.353	0.0944	0	0
1×10^{-4} M	0.447	0.1460	0	0
1×10^{-5} M	0.480	0.1500	0	0
1×10^{-6} M	0.405	0.1260	0	0

TABLE 3. EFFECTS OF HYDROCORTISONE ON 7777 TC CELLS

Concentration	LDH I.U./10^6 cells	MDH I.U./10^6 cells	AFP µg/10^6 cells	*
0 (control)	0.504	0.122	0	2.70×10^6
1×10^{-4} M	1.200	0.431	0	2.94×10^6
1×10^{-5} M	1.000	0.332	0	2.04×10^6

*Mean Cell Number in plate cells/µl.

TABLE 4. EFFECTS OF HYDROCORTISONE ON IOWA 7777 CELLS*

7 days Incubation: Concentration	LDH I.U./10^6 cells	MDH I.U./10^6 cells	AFP μg/10^6 cells	**
0 (Control)	0.186	0.0401	491.4	1.75×10^6
1×10^{-4}	0.556	0.1740	103.8	1.06×10^6
1×10^{-5}	0.515	0.1280	132.5	0.83×10^6

*Average value from triplicate samples.
**Mean Cell Number in Plate

CONTROL OF ONCOFETAL ANTIGEN (ALPHA-FETOPROTEIN) PRODUCTION AS AN EPIPHENOMENON OF TUMORIGENESIS

T.J. Yoo, C. Kuo, S. Patil, U. Kim, L. Ackerman, P. Cancilla, and H. C. Chiu

Departments of Medicine, Pediatrics and Pathology, University of Iowa Hospitals and VA Medical Center, Iowa City, Iowa 52240

INTRODUCTION

Alpha-fetoprotein (AFP) has been measured in the culture medium from mouse hepatoma;[1] and, there are many studies dealing with the relationship between AFP production and proliferation of hepatocyte, AFP production with chromosomal changes and hormonal regulation of AFP production.[2-7] However, the mechanisms regulating AFP synthesis, and AFP's role in regulating tumor growth are not yet fully understood. In our study of tumor antigen we found AFP production by hepatoma culture cells declines on repeated subculturing. Our observation on production of AFP by a few hepatoma lines *in vitro* are described here.

MATERIALS AND METHODS

<u>Hepatoma</u>: Chemically induced rat hepatoma, Morris hepatoma 7777, was maintained by serial subcutaneous passage into Buffalo rats.[8] Both lines of Morris hepatoma 7777 were obtained from Dr. John S. Thompson of the University of Iowa after the 113th "Iowa" generation and the 177th "San Diego" generation respectively.

<u>Culture Medium</u>: Lewis medium, a variant of M-199, was used in this study.

<u>Rat Hepatoma Culture</u>: Rat hepatoma culture was obtained after treating tumor tissue with 0.25% trypsin. Cells were seeded in plastic culture flasks containing Lewis medium.

<u>Preparation of Antiserum</u>: Monospecific anti-rat AFP antiserum was prepared from the antiserum of New Zealand white rabbits. The rabbits were immunized subcutaneously with 10 mg of rat AFP-rich proteins (obtained from DEAE-cellulose chromatography of Morris hepatoma 7777 tumor rat serum). After the antiserum was obtained, it was passed through an immunoadsorbent column conjugated with normal Buffalo rat serum.[10] AFP was obtained by radial immunodiffusion.[11]

<u>Electron Microscopic Studies</u>: Electron microscopic studies were done using cells fed with fresh media for two hours prior to fixation. The media was removed from the culture dishes and replaced with a fixative consisting of 2% glutaraldehyde in 0.16 M sucrose

buffered with 0.1 M cacodylate buffer at pH 7.4. After a rinse in 0.1 M cacodylate buffer, the tissue was processed through 1% osmium tetroxide in cacodylate buffer, uranyl acetate in malate buffer, graded alcohols, and ethanol-epon mixture. Epoxy resin was then poured to a depth of 2 mm and polymerized overnight. A Sieman's Elmiskop 101 electron microscope was used for electron microscopy.

Chromosome Studies: Chromosome studies were done using both Q- and G-banding procedures.

RESULTS

AFP production *in vitro* and *in vivo* by Morris hepatoma 7777:

The culture medium used for AFP determinations was collected from confluent cultures after one week of growth without changing media. The culture medium was tested for AFP by immunodiffusion and immunoelectrophoresis. The amount of AFP in the culture medium is shown in Table 1.

At first the cultivated rat hepatoma tumor cells produced AFP, but the ability to produce detectable amounts diminished as subculturing progressed. After eight months of *in vitro* cultivation the established tumor cell completely stopped detectable production.

Due to the AFP production loss *in vitro*, we followed the serum AFP levels of Buffalo rats. They were given intramuscular injections of 1×10^6 rat hepatoma cells that had been cultured for eight months.

A solid tumor showing significantly lower AFP production levels than the *in vivo* stock tumor line formed three weeks after the injections. An average 0.11 mg AFP/ml in tissue cultures of six tumor rats resulted--compared with an average 5.14 mg AFP/ml of serum measured at the same tumor size in five stock tumor rats. No significant change in serum AFP was noted during the experiment. The usual serum AFP level in terminal tumor bearing rats in our laboratory is 10-15 mg AFP/ml.

Table 2 shows serum AFP levels in tissue culture tumor rats and stock tumor rats at 100 cm^3 tumor size.

Fourteen months after the initiation of *in vitro* culture, tumors were induced in 15 Buffalo rats with cultured Morris hepatoma 7777 cells. The Mancini method detected no AFP production at 14, 27, and 40 days after tumor implantation. Four 7777 TC animals with tumor induced by 19-month *in vitro* cultures were shown to have no detectable AFP levels. Tumor cell culture from one of the non-AFP-producing tumors was checked for AFP production. The first subculture medium (1×10^6 cell/ml) failed to show AFP, even after tenfold concentration by Amicon PM-30 membrane.

Detection to AFP in/on the hepatoma cells: Two strains of Morris hepatoma 7777 cells were grown on coverslips and tested for AFP synthesis by immunofluorescence. The first strain, 7777 San Diego, was freshly derived (thirty days *in vitro* culture) from a tumor bearing rat. This tumor line had been maintained continuously *in vivo* by serial transplantation in female Buffalo rats. This shows positive AFP in these cells.

The second strain, 7777 TC, was derived from a tumor that had been induced 62 days earlier by an intramuscular injection of 10^6 viable cultured tumor cells. The cell culture had spent eight months *in vitro* after isolation from the same *in vivo* tumor line as the 7777 San Diego culture. The 7777 TC showed negative fluorescence, whereas 7777 San Diego showed positive fluorescence.

Electron Microscope Study: The comparison of cellular organelles of the two hepatoma cells (AFP producing 7777 Iowa, and the variant non-AFP-producing 7777 TC) are shown in figures 1 and 2.

In each line the nuclei were characterized by a prominent nucleus, dispersed chromatin and smooth to slightly irregular nuclear membranes. The cytoplasmic membranes had microvillar projections. Occasionally cell junctions were present when cells were in close apposition. The cytoplasm contained prominent endoplasmic reticulum and often the endoplasmic reticulum was closely applied to mitochondrion. Electron dense bodies and membraneous inclusions were variable findings.

Karyotypic Analysis: At least fifty cells of each cell type were analyzed with trypsin Giemsa banding technique (Figures 3-5).

A. 7777 Iowa--
When the initial chromosome studies were performed with unbanded procedure, the tetraploidy condition was predominant without apparent structural rearrangement. Ten months later, these cells had undergone numerical and structural alterations. The chromosome range was 45-155. There was marked aneuploidy. Twenty-two percent of the cells were in the hypertriploid range.

The most prevalent markers seen were: (1) a small amount of extra material in the short arm of chromosome 1 in 70% of the cells;

(2) a small amount of extra material in the short arm of chromosome 3 in 20% of the cells; (3) a metacentric chromosome which was probably originated by the fusion of two chromosomes 11 in 20% of the cells; and (4) a marker chromosome resulting from the fusion of chromosomes 3 and 4 in 20% of the cells.

B. 7777 TC--

Initial karyotypic analysis showed that the majority of the cells were in the diploid range with numerous structural rearrangements such as centric fusions and dicentrics. A very few normal chromosomes were present. After six months, the chromosome range was 42-92. Nearly 40% of the cells had chromosomes in the diploid range (42-44) and another 40% in the near tetraploid (75-84).

The most common chromosome markers found were: (1) a translocation chromosome involving two chromosomes 4 in 70% of the cells; (2) a chromosome 1 marker similar to the one found in Iowa 7777 in 14% of the cells; and (3) a chromosome 7 marker seen in 8% of the cells.

C. 7777 San Diego--

The majority of the cells (58%) contained chromosomes in the near diploid to diploid level (35-44). Aneuploidy conditions were less marked in these cells. Only 8% were in the near tetraploid range (75-84).

The most consistent single marker chromosome was chromosome 7. No other obvious markers were seen as in 7777 Iowa or 7777 TC.

DISCUSSION

From the beginning of the first cell culture to the end of the eight month subculture (and in on-going cultures) we have not seen any significant changes in cell morphology, nor was there change in microscopic histology section or in electron microscopic examination. However, the early cultures produced and released AFP during *in vitro* cultivation, while AFP production declined upon continuous subculture. It has been shown that in the course of *in vitro* cultures, chemically induced mouse hepatomas either ceased to produce αf globulin while albumin and β-globulin were present in the medium, or they lost the ability to produce not only the embryonal α-globulin, but serum albumin as well.[1] This contrasted with our findings with Morris hepatoma 7777 TC. Again, in contrast to previous reports, our tumor cell line did not decrease transplantability *in vivo* after long term culture.[1] Non-AFP-producing cells grow as well as AFP producing cells *in vivo*. There were no appreciable differences in the proliferation rates *in vitro* or *in vivo* between the original 7777 cultures and those maintained longer in culture. The growth rate of 7777 TC *in vivo* is similar to the growth rate of 7777 San Diego and 7777 Iowa. Yet, 7777 TC failed to produce AFP. Therefore, AFP production is not directly related to proliferation.

Becker, et al., using transplantable rat hepatoma, found a reverse relationship between karyotype and AFP production.[12] Thus, rapidly growing tumors with aneuploid number demonstrate intense production of normal proteins and AFP, while near diploid tumors demonstrate little or none. This study was extended by Sell and Morris.[6] With many lines of Morris hepatoma, AFP production was more increased in the aneuploid with structural changes than in the aneuploid without structural changes of the chromosome, and these aneuploid tumors produced more AFP than did the diploid. Furthermore, they demonstrated that slow-growing, well differentiated tumors generally do not produce elevated serum AFP concentrations, while fast-growing, poorly differentiated tumors produce high serum AFP concentrations.

However, fast-growing near-diploid tumors did not produce significant AFP serum levels, while slow-growing aneuploid tumors may. They concluded that the potential of a given tumor line to produce AFP appears to be controlled genetically, but the degree of expression of this potential is dependent on the tumor growth rate.[6] In view of this, results of chromosome studies in our lines are of interest. First, they support the theory that aneuploid tumors produce excessive AFP. Both lines stem from the same original cells. However, though no apparent changes in the rate of proliferation and differentiation occur, the degree of AFP production is different. The 7777 TC with nearly 50% of the cells in the diploid range, with many structural aberrations, failed to produce AFP; whereas the 7777 Iowa, with predominantly tetraploid cells, did not lose the potential for AFP production. Although this line now had marked aneuploidy with certain structural aberrations, it still produced AFP.

Detailed chromosome analyses of several transplantable hepatomas have been reported with banding procedures.[7,13] Tumors with a longer transplant history showed the most complex non-random structural rearrangement.[13] Marker chromosomes 2, 6, and 10 were derived by fusion and/or pericentric inversions. Though a pericentric inversion of chromosome 2 was present in five out of six hepatomas they studied, no rearrangement common to all hepatomas was found.

A marker 7 was seen in the Morris hepatoma 7777[7] leading to speculation on the possible association of this marker and AFP production.[6]

The apparent loss of AFP production and the disappearance of the chromosome 7 marker in our TC 7777 line is of interest. It is likely that the chromosome 7 marker may indeed be associated with AFP production as has been postulated by Wolman, et al.[7] However, that raises the questions concerning the 7777 Iowa line which produces AFP but lacks this particular chromosome marker. In these cells, the chromosome 1 marker is the most common. Whether the additional material present on the short arm is the same as on chromosome 7 has not been determined. Our initial impression is that it is different. Additional banding studies such as R (Reversed) and T (Telomeric) would aid in determining the nature and origin of the extra material on chromosomes 7 and 1 and its significance for association with AFP production.

Thus it appears that although AFP is an oncofetal antigen and a gene product of tumor cells, the control mechanism for AFP production could be an epiphenomenon of the tumor cells and not related to the control of tumor growth and differentiation.

BIBLIOGRAPHY

1. Irlin, J.S., Perova, S.D., Ablev, G.J. Changes in the Biological and Biochemical Properties of the Mouse Hepatoma During Long-Term Culture *In Vitro*. Intl. J. Cancer 1:337-347, 1966.
2. Leffert, H.L., Koch, K.S., and Tubakava, B. Present Paradoxes in the Environmental Control of Hepatic Proliferation. Cancer Res., 36:4250-4255, 1976.
3. Leffert, H.L., and Sell, S. Alpha Fetoprotein Biosynthesis During the Growth of Differentiated Fetal Rat Hepatocytes in Primary Monolayer Culture. J. Cell Biol., 61:8-15, 1976.
4. deNechaud, B., DeBecker, J.E., and Potter, V.R. Effect of Glucocorticoids in Fetoprotein Production by an Established Cell Line Morris Hepatoma 8994. Biochem. and Biophys. Res. Comm., 68:8-15, 1976.
5. Sell, S., Becker, F.F., Leffert, H.L., Watabe, H. Expression of an Oncodevelopmental Gene Product (Alpha Fetoprotein) During Fetal Development and Adult Oncogenesis. Cancer Res., 36:4239-4249, 1976.
6. Sell, S., and Morris, H.P. Relationship of Rat Alpha Fetoprotein to Growth Rate and Chromosome Composition of Morris Hepatoma. Cancer Res., 34:1413-1416, 1974.
7. Wolman, S.R., Cohen, T.I., and Becker, F.F. Chromosome Analysis of Hepatocellular Carcinoma 7777 and Correlation with Alpha Fetoprotein Production. Cancer Res., 37:2624-2627, 1977.
8. Morris, H.P., and Wagner, B.P. Induction and Transplantation of Rat Hepatomas with Different Growth Rates (Including "Minimal Deviation Hepatoms"). *In* Methods in Cancer Research, 4:125-152, 1968.

9. Lewis, L.J., Hoak, J.C., Maca, R.D., and Fry, G.L. Replication of Human Endothelial Cells in Culture. Science, 181: 453, 1973.
10. Cuatrecases, P. Protein Purification by Affinity Chromatoggraphy: Derivation of Agarose and Polyacrylamide Beads. J. Biol. Chem., 245:3059, 1970.
11. Mancini, G., Carbonara, A.D., Heremans, J.F. Immunochemical Quantitation of Antigens by Single Radial Immunodiffusion. Immunochem., 2:235, 1965.
12. Becker, F.F., Wolman, S.R., Asofsky, R., and Sell, S. Sequential Analysis of Transplantable Hepatocellular Carcinomas. Cancer Res., 35:3021-3026, 1975.
13. Kovi, E., and Morris, H.P., Chromosome Banding Studies of Several Transplantable Hepatomas. Adv. Enz. Reg., 14:139-162, 1976.

TABLE 1. THE CHANGE OF α-FETOPROTEIN PRODUCTION IN THE ESTABLISHED RAT HEPATOMA CULTURES

CULTURE SAMPLE DATE		AFP LEVELS (mg/ml)
5/30/75	(no concentration)	0.021
5/30/75	(fivefold concentration)	0.180
9/15/75	(no concentration)	not detectable
9/15/75	(fivefold concentration)	0.030
9/15/75	(twenty-five fold concentration)	0.172
11/10/75	(tenfold concentration)	not detectable
1/23/76	(fivefold concentration)	not detectable
1/23/76	(twenty-five fold concentration)	not detectable

TABLE 2. SERUM AFP LEVELS IN THE *IN VIVO* STOCK TUMOR LINE TUMORS AND TISSUE CULTURE TUMORS

Tumor Rat Individuals		Serum AFP levels (mg/ml)	t test
	a	4.75	
	b	4.05	
Stock Tumor Line	c	6.25	(5.13 ± 0.81)†
	d	5.30	
	e	5.30	
			$p < 0.01$
	a	0.086	
	b	0.086	
Tissue Culture Tumor Line*	c	0.129	(0.11 ± 0.04)†
	d	0.172	
	e	0.086	
	f	0.086	

*Tumors are induced by tumor culture cells derived from the stock tumor

†Mean ± standard deviation.

Figure 1. Electron micrograph of 7777 TC cells with prominent endoplasmic reticulum, irregular nuclei and cytoplasmic contacts between cells (4600 x).

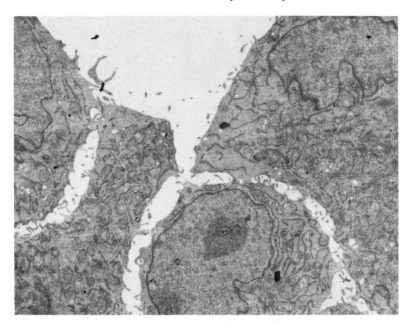

Figure 2. Electron micrograph of 7777 Iowa cells with mitochondria and rough endoplasmic reticulum in close apposition, scattered dense bodies and a phagocytic vacuole containing membranous debris (7600 x).

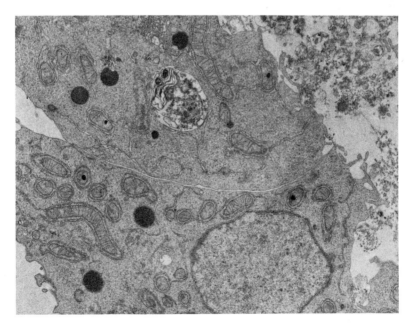

Figure 3. A G-banded karyotype of 7777 Iowa with 61 chromosomes showing a marker chromosome 1 (arrow, top row) and a metacentric chromosome resulting from fusion of two chromosomes 11 (arrow, second row).

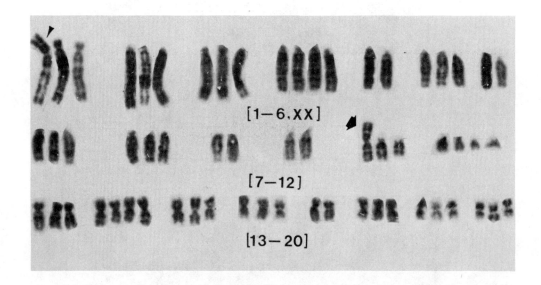

Figure 4. A G-banded karyotype of 7777 TC with 44 chromosomes showing a large metacentric chromosome (arrow).

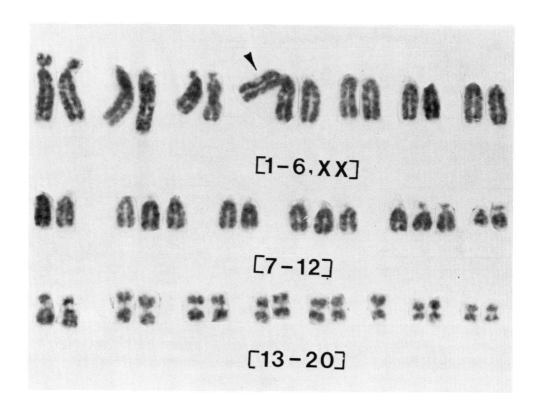

Figure 5. A G-banded karyotype of 7777 SD with 44 chromosomes showing a marker chromosome 7 (arrow).

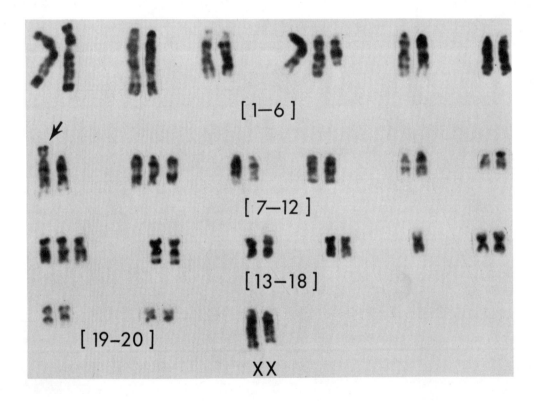

ENZYMES AND ISOZYMES IN HUMAN HEPATOMA

Doris Balinsky, Eftihia Cayanis, Kathryn Hammond and Roger Cummins

The South African Institute of Medical Research, Johannesburg, South Africa (Present address: Department of Biochemistry, Iowa State University, Ames, Iowa)

Human primary hepatoma occurs with high incidence in southern Africa, especially in parts of Moazmbique, where 1:1000 of the population die of this disease. Detailed studies of altered enzyme activities (6) and isozyme patterns (5) in rat hepatomas of varying growth rate have been made. We considered it of interest to examine the levels and isozyme patterns of selected enzymes in the human hepatomas in order to determine what metabolic changes occur in this disease. Two pathways in particular were considered of interest, *viz.* those of DNA synthesis and of carbohydrate metabolism. Since adult liver does not divide under normal conditions, enzymes of DNA synthesis might be expected to increase in tumors which represent dividing tissue.

MATERIALS AND METHODS

We assayed five enzymes of DNA synthesis, three of these being enzymes of thymidine metabolism, since thymidine is unique to DNA (Figure 1). Autopsy tissue, obtained within one to six hours of death, was used. The enzyme levels were compared in normal adult liver (obtained from patients who had no liver disease), in "host" liver (from patients who had hepatomas), in hepatomas and in fetal human liver (representing growing, dividing human liver).

RESULTS

The levels of thymidylate kinase, thymidylate synthetase, DNA polymerase and deoxycytidylate deaminase were all elevated in the human hepatoma relative to those in the host liver of the same patient, but were, in general, considerably lower than those in fetal liver except in the case of deoxycytidylate deaminase. Thymidine kinase levels in the hepatoma were not elevated; this is of interest because of the use of 5-fluorouridine as a carcinostatic agent. This compound is utilized as a substrate by thymidine kinase to form 5-fluorouridylate, which inhibits thymidylate synthetase. However, if thymidine kinase is not elevated, 5-fluorouridine would probably not be an effective drug.

The liver is very active in carbohydrate metabolism, hence we examined many enzymes of these pathways (1,3). Figure 2 shows the overall pathways of carbohydrate metabolism. Glycolysis, the breakdown of glucose to give, ultimately, lactate, is common to all tissues. It might be expected to increase in growing tissues, since this is an energy-producing pathway, providing the ATP required for many biosynthetic processes. Gluconeogenesis, which is virtually the reverse of glycolysis, is unique to liver and kidney cortex, hence is of interest to examine in hepatome. It will be seen in Figure 2 that there are certain key steps where a different enzyme participates in glycolysis from that catalyzing the reverse pathway of gluconeogenesis. We assayed all the enzymes specific to gluconeogenesis and found them all to be considerably reduced in the hepatoma (1,3). Figure 3 shows the data obtained for pyruvate carboxylase, phosphoenolpyruvate carboxykinase and glucose 6-phosphatase. A line is used to join the activity in the tumor with that in the corresponding uninvolved liver from the same patient. We can conclude that, in the hepatoma, we have lost the liver-specific enzymes characteristic of differentiated adult liver tissue.

The three key enzymes of glycolysis--hexokinase, phosphofructokinase and pyruvate kinase, were also assayed. Only pyruvate kinase was consistently elevated, although only to approximately twice the level of the host liver. Figure 4 shows the data for hexokinase, the first enzyme of the glycolytic pathway. In a 3'-methyldimethylaminoazobenzene-induced, transplantable rat hepatoma, the levels in the hepatoma were consistently higher than in the host liver. However, in humans the patterns were not so clearcut, and the levels were not consistently different in hepatomas and host livers.

Enzyme activities, while very useful to give an indicating of metabolic change, only give part of the story. A more clearcut picture of the loss of tissue-specific enzymes is given if one looks at isozyme patterns. Isozymes are defined as different forms of an enzyme which catalyze the same reaction but differ in various physical and/or physicochemical properties. The physical property most easily accessible to experimental observation is electric charge. Proteins subjected to electrophoresis at a particular pH will migrate toward the anode or cathode depending on their electric charge. By staining specifically for the enzyme activity, one can localize the position of the isozymes (1,4).

Figure 5 shows the isozymes of hexokinase in normal and cancerous human liver, muscle and hepatoma cells in culture. When stained with 0.5 mM glucose (left half of gel), both normal and host liver showed two bands (HK I and HK III); hepatoma showed reduction of HK III and appearance of HK II. This showed more clearly in the hepatoma cell culture where HK III was absent and HK II very prominent. Fetal liver showed a similar pattern to hepatoma, with an additional band, HK IA. When staining with high glucose (0.1 M), (right half of gel), HK IV, or glucokinase was seen in the surgically removed host liver, but not in the corresponding hepatoma. HK IV is unstable to freezing, hence was not seen in the normal liver which had been stored frozen. HK III is inhibited by high glucose, hence is not visible in these gels. However, an additional band, which we designated HK I II$_f$, was seen in all samples except muscle, when the gels were run under the conditions described. This was not observed in blank gels stained in the absence of ATP or glucose, and was observed whether EDTA was present in the electrophoresis buffers or not. The disappearance of HK III and HK IV, the two liver-specific isozymes in the hepatoma is another example of loss of liver-specific enzymes in this tissue. However, HK III was retained in some cases, and HK II did not always occur in human hepatoma. It is of interest that the human host liver frequently showed a pattern intermediate between that of normal liver and hepatoma, with some HK II present, indicating that the tumor may be affecting the isozyme pattern of the adjacent liver.

DISCUSSION

As mentioned above, pyruvate kinase was the one key glycolytic enzyme whose activity was increased in hepatoma relative to the activity in host liver. The pyruvate kinase isozyme patterns were examined. Adult human liver had mainly PK-L and occasionally had a trace of PK-M$_2$. The patterns in both hepatomas and host livers were rather variable as shown in Figure 6. In general, compared to normal liver, there was decrease of PK-L and increase of PK-M$_2$. Since the total PK activity was increased, this represented a large increase of PK-M$_2$ relative to normal liver. PK-M$_1$, the isozyme predominating in adult muscle, was also found in some hepatomas, PK-L was sometimes retained, and intermediate bands between PK-L and PK-M$_2$, possibly representing hybrids composed of both L and M$_2$ subunits, were observed in some cases. The altered pattern in the host liver could be due to humoral factors put out by the hepatoma; also, since PK-L is under hormonal and nutritional control, its relatively low proportion in host liver could be due to the poor nutritional state of these terminally ill patients. The change of isozyme pattern is significant because PK-L is a highly regulated enzyme, being an allosteric enzyme subject to feedforward activation by fructose 1,6-bisphosphate and to inhibition by ATP and alanine. Human tumor PK-M$_2$, on the other hand, is not allosteric, is not activated by fructose 1,6-bisphosphate and is less inhibited by ATP (Balinsky, et al., 1973b).

Thus in human hepatoma we see loss of enzymes and isozymes specific to adult differentiated liver and appearance of enzymes which are less regulated.

REFERENCES

1. Balinsky, D., Cayanis, E., and Bersohn, I. (1973a) Cancer Research 33:249-255.

2. Balinsky, D., Cayanis, E., and Bersohn, I. (1973b) International Journal of Biochemistry 4:489-501.

3. Hammond, K.D. and Balinsky, D. (1978a) Cancer Research 38: 1317-1322.

4. Hammond, K.D. and Balinsky, D. (1978b) Cancer Research 38: 1323-1328.

5. Schapira, F. (1973) Advances in Cancer Research 18:77-153.

6. Weber, G. (1972) Gann Monograph 13:47-77.

Figure 1

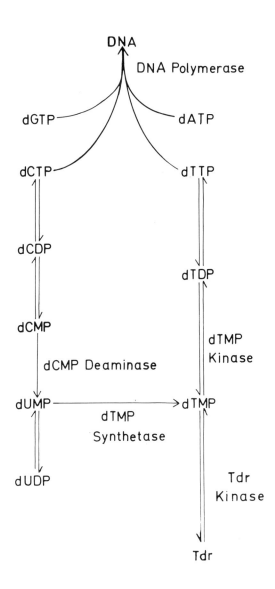

BALINSKY

Figure 2

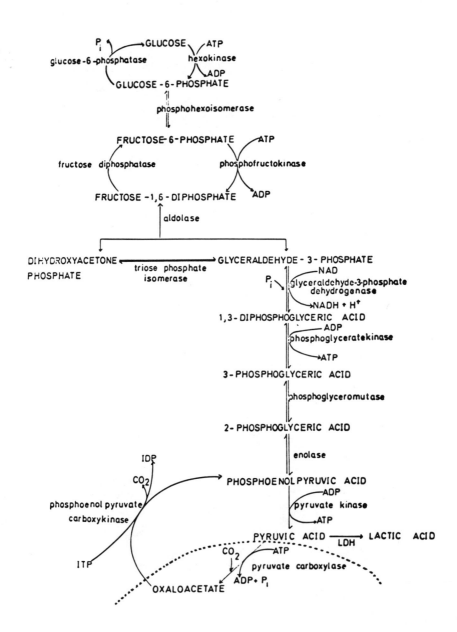

Figure 3

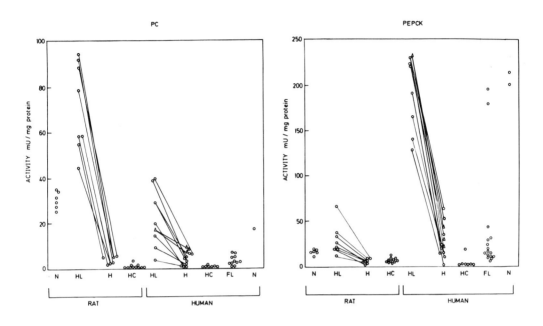
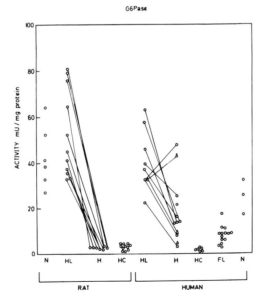

Figure 4

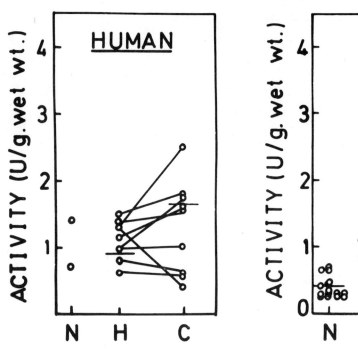

Figure 5

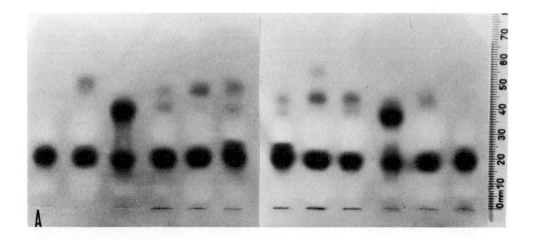

Figure 6

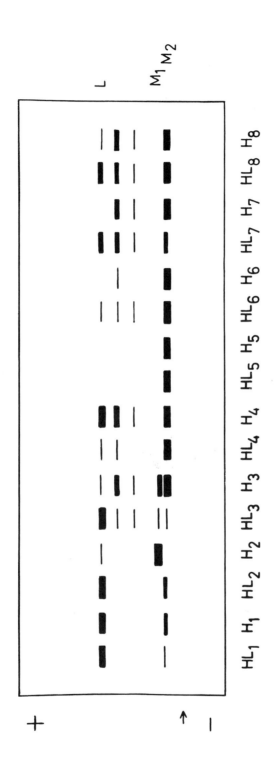

THE ASSOCIATION OF AN ULTRAFAST ALKALINE PHOSPHATASE ISOENZYME WITH MALIGNANCY

John Koett, M.D.,[1] James Howell, M.D.,[1] and Paul L. Wolf, M.D.[2]

[1]*Department of Laboratory Medicine, U. S. Naval Regional Medical Center, San Diego, California, and* [2]*Professor of Pathology, Department of Pathology, University of California, San Diego, School of Medicine, La Jolla, California*

INTRODUCTION

Recently we have studied five patients who have exhibited an unusual alkaline phosphatase isoenzyme (ALP E.C. No. 3.1.3.1.) migrating in an ultrafast position electrophoretically on cellulose acetate. This ALP isoenzyme has been identified in patients with benign and malignant liver diseases. In addition, a number of these patients exhibited a regular ALP liver isoenzyme and a fast (preliver) ALP isoenzyme in conjunction with the ultrafast ALP liver isoenzyme.

MATERIALS AND METHODS

Patients' sera were selected on the basis of an elevated ALP which were electrophoresed immediately or stored at $-70^\circ C$ until the day of electrophoresis. Clinical information was correlated with the laboratory findings.

The isoenzymes of alkaline phosphatase were delineated using the Helena Laboratory alkaline phosphatase isoenzyme procedure which entailed the use of Titon III-Iso cellulose acetate plates, 60 x 75 mm, Electro HR (barbital) buffer, and alpha naphthol ASMX reagent which was subsequently coupled with the diazo compound, Fast blue RR, yielding sharp blue visible bands. The developed plates were then scanned on an Auto Scanner Flur-Vis densitometer.

Identification of individual isoenzyme bands was made utilizing reference alkaline phosphatase control sera supplied by the Helena Laboratory consisting of bovine, intestinal and liver enzymes.

RESULTS

We observed a fast liver fraction in conjunction with an ultrafast liver band in three patients with various diseases including

---

[1]The opinions or assertions herein are the private views of the authors and are not to be construed as official or as reflecting the views of the Department of Defense or the Department of the Navy.

carcinoma of the head of the pancreas with metastasis to the liver (Figure 2), ovarian carcinoma with suspected metastasis to the liver (Figure 3) and breast carcinoma metastatic to the liver (Figure 4). This unique ultrafast liver isoenzyme has also been observed in conjunction with a regular liver isoenzyme in the absence of the fast liver fraction (Figures 1,5,6). The representative combination of isoenzymes with correlated clinical histories are listed in Table I. Representative densitometry tracings from cases 1 through 6 and cases 8, 9, 10 and 14 are illustrated. In all patients there was an elevation of the serum ALP but not necessarily an elevation of the patient's bilirubin. The total ALP ranged from 95 U/L to approximately 600 U/L as determined on the SMAC instrument (Technicon Inc., Tarrytown, New York). Low levels of total ALP did not appear to preclude the presence of the ultrafast liver isoenzyme ALP as indicated in patient M.D. (Figure 5) whose total activity was 95 U/L. Multiple determinations of the isoenzyme electrophoretic pattern did not alter the presence or location of the isoenzyme even after storage at -70°C for approximately one to two weeks. Followup studies on the ultrafast fraction in patients with chronic disease continued to demonstrate the presence of this ultrafast fraction as was the case with patient G.W. (Figure 1) after a one month lapse between analyses. In contrast, patient H.R. (case 7) listed in Table I with a transient viral syndrome demonstrated an elevated ALP and LD which on fractionation of the former enzyme demonstrated a fast liver isoenzyme that disappeared on resolution of the viremia.

DISCUSSION

This paper reports the finding of a unique isoenzyme of ALP which has not been previously observed on review of the literature. Since this isoenzyme migrates toward the anode and is faster than the isoenzyme described by Fritsche and Adams-Port (1972), we have chosen to call it the ultrfast liver isoenzyme of ALP. Previous investigators have identified a regular liver isoenzyme and an abnormal anodally migrating enzyme referred to as pre-liver (Fritsche and Adams-Port, 1972 and 1974) or fast liver. In contrast to these ALP isoenzymes which we have likewise observed, the ultrafast isoenzyme migrates faster toward the anode than the other two liver isoenzymes in the area corresponding to albumin on cellulose acetate electrophoretic media.

The ultrafast liver fraction was primarily observed in patients with malignancy associated with extrahepatic and/or intrahepatic cholestasis. In addition, patients with Laennec's cirrhosis and alcoholic hepatitis also exhibited the ultrafast isoenzyme. A number of patients with malignancy had the simultaneous presence of three liver isoenzymes: the regular enzyme, fast liver and the ultrafast liver isoenzyme. This pattern did not occur in patients with cirrhosis or alcoholic hepatitis. One patient had an ultrafast liver fraction caused by a drug-induced hepatitis (Case 6, Figure 6). We have also observed a fast liver fraction associated with Hodgkin's disease causing a large filling defect of the liver on liver-spleen scan (Case 8, Figure 7). We have also seen this similar isoenzyme pattern in one patient with hemochromatosis (Case 9), renal cell adenocarcinoma (Case 10), and another patient with disseminated coccidioidomycosis (Case 11).

We have recently become disenchanted with the heat stability test for ALP since we do not feel this test yields definitive, clinically useful information. Confusing results are obtained with the heat test due to the presence of multiple isoenzymes of ALP exhibiting different heat inactivation attributes (Fishman, 1974; and Whitby and Moss, 1975). We therefore have sought another modality to aid in the differential diagnosis of an elevated serum ALP. We chose the Helena Laboratory isoenzyme electrophoresis procedure on cellulose acetate membranes because of the ease of handling, stability of the supporting membrane and media and the reproducibility of the results. All these factors have made this procedure attractive for the laboratorian in an extremely active clinical laboratory. In contrast, although acrylamide gel electrophoresis is more discriminating because of its molecular sieving action as well as its separation on the basis of charge, this media requires preparation of individual components and polymerization steps and, in some cases, pre-electrophoresis to remove any potential interfering substances that might inhibit enzymatic activity (Gordon, 1971).

Previous investigators using different separating media have identified the presence of a fast liver band in patients with liver disease (Fritsche and Adams-Port, 1974; and Afinja and Baron, 1974), and this band may represent the slow moving isoenzyme noted in polyacrylamide gel electrophoresis (Fritsche and Adams-Port, 1974; and Price and Simmins, 1974). DeBroe, et al., (1975) found that the high molecular weight alkaline phosphatase with rapid migration on cellulose acetate electrophoresis corresponded to a membrane-vesicle fraction seen with electron microscopy in patients with cholestasis. The unique isoenzyme which we have observed may be a post-translationally modified enzyme of the liver type associated with a lecithin-liver isoenzyme complex which induces a change in its electrophoretic mobility as described by Price and Simmins (1974) in obstructive liver disease. However, studies in our laboratory utilizing Triton X-100 anionic detergent induced no appreciable change in the isoenzyme pattern previously noted. At the present time, we are not able to identify the genesis of this newly described ALP isoenzyme. We are presently actively investigating the inhibitory effect of different amino acids as well as classical chemical agents utilized to disrupt hydrogen bonding to further characterize the ultrafast isoenzyme. Moreover, we are investigating the effect of enzymatic degradiation of the new isoenzyme in an attempt to define its reactivity in the absence of potentially critical carbohydrate moieties.

Previous reports have identified three isoenzymes of alkaline phosphatase in patients with cirrhosis (Fishman, 1974). They include the regular liver, bone and intestinal isoenzyme. The elevation of the intestinal isoenzyme has been noted by us (Table 1) and may be related to anoxic injury to the gastrointestinal tract induced by portal hypertension. The elevated bone fraction has been believed to be secondary to a defect or decreased activity in the 25-hydroxylase enzyme of the liver in cirrhosis, which converts cholecalciferol to its 25-hydroxylated product, which is the precursor to the active metabolite 1,25 di-hydroxycholecalciferol via the one hydroxylase of the kidney. This latter Vitamin D derivative has been demonstrated to be intimately involved with calcium absorption from the gastrointestinal tract. With inadequate physiological quantities, a relative hypocalcemia ensues with secondary activation of PTH secretion by the parathyroid glands thereby increasing osteoblastic and osteoclastic activity with resultant elevation of the bone isoenzyme of ALP. We have seen this isoenzyme pattern in a number of patients with alcoholic cirrhosis.

Recently, (Ehrmeyer et al., 1978) demonstrated the presence of a fast-liver fraction referred to as fast, homoarginine sensitive alkaline phosphatase that migrate in the α_1 serum protein position on cellulose acetate media in patients with various carcinomas. Unlike Ehrmeyer's findings, our ultra-fast liver fraction migrates anodally to his identified band. However, in concert with Ehrmeyer's findings, we did identify a pre-liver or fast-liver fraction in patients with cancer.

Recently the ultrafast liver band has been attributed to an artifact caused by a bilirubin albumin complex and Fast Blue RR utilized in the electrophoretic technique (Harding, 1978). However, another publication (Tsung, 1978) disputed Harding's conclusion and described two patients with normal serum bilirubin levels; one with granulomatous hepatitis and the other with carcinoma of the pancreas metastatic to the liver who demonstrated an ultrafast alkaline phosphatase band on electrophoresis. Because of these reports, we chose random serum samples from patients with variably elevated serum bilirubin and alkaline phosphatase.

To determine if the ultrafast liver isoenzyme was an artifact of a complex of bilirubin and albumin and Fast Blue RR, a number of serum samples with elevated serum bilirubin and alkaline phosphatase levels were electrophoresed and subsequently developed with ASMX (containing alpha naphthol phosphate) and Fast Blue RR or barbital buffer (pH 8.8 with Fast Blue RR only. Two of the specimens studied were from patients lited in Table 1 (cases 15 and 16) and are indicated in Table 2. The remaining specimens were chosen randomly without regard to their clinical histories and solely on the basis of an elevated serum bilirubin.

The results listed in Table 2 (patients 15 and 16) demonstrate that the "ultrafast" liver isoenzyme in these two patients is really an artifactual band produced by the interaction of bilirubin, albumin and Fast Blue RR as suggested by Harding. However the results in patients 3 through 5 are not consistent with Harding's findings and are suggestive of the presence of a true alkaline phosphatase isoenzyme migrating as an ultrafast liver isoenzyme supporting Tsung's report. In addition in these patients (3 through 5) a band that was displaced anodally to the ultrafast band and very small in amplitude in comparison to the ultrafast band was observed. This small absorbing band probably represents a bilirubin albumin complex.

In conclusion, the ultrafast alkaline phosphatase isoenzyme appears to be a true isoenzyme and associated most commonly with neoplastic disease of the liver.

In summary, we have described a new isoenzyme of alkaline phosphatase which surprisingly has not been previously reported. Whether or not this isoenzyme is only demonstrated utilizing the alpha naphthol substrate of the Helena Laboratory procedure has not been elucidated to date. Perhaps other substrates such as thymolphthalein phosphate, beta glycerol phosphate, or indoxyl phosphate will likewise react with this enzyme. We hope this paper will initiate other investigations to help identify and corroborate our findings and attempt to assign this isoenzyme clinical and diagnostic relevancy.

REFERENCES

1. Afinja, A.D., and Baron, D.N. (1974) Plasma alkaline phosphatase isoenzymes in hepatobiliary disease. J. Clin. Path. 27, 916-920.
2. DeBroe, M.E., Borgers, M., and Wieme, R.J. (1975). The separation and characterization of liver plasma membrane fragments circulating in the blood of patients with cholestasis. Clinical Chimica Acta 59, 369-372.
3. Ehrmeyer, S.L., Joiner, B.L., Kahan, L., Larson, F.C., Metzenberg, R.L. (1978) A cancer-associated, fast, homoarginine-

sensitive electrophoretic pattern of serum alkaline phosphatase. *Cancer Research* 38,
4. Emery, A.J., Jr., and Dounce, A.L. (1955) Intracellular distribution of alkaline phosphatase in rat liver cells. *J. Biophys. Biochem. Cytol.* 1, 315.
5. Fishman, W.H. (May, 1974). Perspectives on alkaline phosphatase isoenzymes. *Am. J. of Med.* 56, 617-650.
6. Fritsche, H.A. Jr. and Adams-Port, H.R. (1972) Cellulose acetate electrophoresis of alkaline phosphatase in human serum and tissue. *Clin. Chem.* Vol. 18, No. 5.
7. Fritsche, H.A. Jr. and Adams-Port, H.R. (1974) High molecular weight isoenzymes of alkaline phosphatase in human sera demonstration by cellulose acetate electrophoresis and physico-chemical characterization. *Clinica Chimica Acta* 52, 81-89.
8. Gordon, A.H. (1971). 3rd Ediction, *Electrophoresis of Proteins in Polyacrylamide and Starch Gel.* North-Holland Publishing Company, Amsterdam-London.
9. Hardin, E., et al. (1978) *Clin. Chem.* 24, 178 (Letter)
10. Price, C.P. and Simmins, H.G. (1974). The nature of the serum alkaline phosphatases in liver disease. *J. Clin. Path.* 27, 392-398.
11. Thines-Sempoux, D. (1973). A comparison between the lysosomal and the plasma membranes. Chapter II. In *Lysosomes in Biology and Pathology*, edited by J. T. Dingg. North-Holland Publishing Company, Amsterdam.
12. Tsung, Swei H., (1978) *Clin. Chem.* 24, 2068, (Letter)
13. Whitby, L.G. and Moss, D.W. (1975). Analysis of heat inactivation curves of alkaline phosphatase isoenzyme in serum. *Clinica Chimica Acta* 59, 361-367.

TABLE I.

Case Number (with figure reference)	CLINICAL/PATHOLOGICAL FINDINGS
1. (Figure 1)	61 year old female with cirrhosis
2. (Figure 2)	72 year old female with carcinoma of the head of the pancreas with metastasis to the liver
3. (Figure 3)	60 year old female with metastatic ovarian carcinoma
4. (Figure 4)	60 year old female with metastatic breast carcinoma
5. (Figure 5)	55 year old female with liver-spleen scan consistent with cirrhosis/multiple areas of isotope uptake on bone scan
6. (Figure 6)	55 year old female with abnormal liver function tests believed to be secondary to allopurinol and/or colchicine--resolution on withdrawal of drugs
7.	30 year old male--viremia with bronchitis
8. (Figure 7)	31 year old male with Stage IV Hodgkin's disease--filling defect on liver-spleen scan
9. (Figure 8)	67 year old female with hemochromatosis--discrete filling defect on liver-spleen scan
10. (Figure 9)	52 year old male with renal adenocarcinoma
11.	23 year old male with disseminated coccidioidomycosis
12.	70 year old male with cirrhosis and alcoholic hepatitis
13.	60 year old male with cirrhosis and portal hypertension
14. (Figure 10)	60 year old male with metastatic adenocarcinoma
15.	58 year old male with metastatic rectal Ca to the liver, duodenum, lung and parietal pleura
16.	19 year old male with hepatitis

ULTRAFAST ALP

ISOENZYMES FRACTIONATION

Intestine	Regan	Bone	Liver	Pre-Liver	Ultrafast liver
+	−	−	+	−	+
−	−	−	+	+	+
−	−	−	+	+	+
−	−	−	+	+	+
−	−	+	+	−	−
−	−	+	+	−	+
−	−	−	+	+	−
−	−	−	+	+	−
−	−	−	+	+	−
−	−	−	+	+	−
−	−	−	+	+	−
−	−	−	+	+	−
+	−	−	+	−	−
−	−	−	+	+	−
−	−	−	+	+	+
−	−	−	+	+	+

Table 2. ISOENZYME

		Liv.	Pre-Liv.	UF* Liv.	Bili mg/dl	Total ALP (U/L)
1.	Case 16, Table 1) Alpha Naphthol Phosphate + Fast Blue RR	+	+	+	>20	247
	Fast Blue RR only	−	−	+		
2.	(Case 15, Table 1) Alpha Naphthol Phosphate + Fast Blue RR	+	+	+	2.9	496
	Fast Blue RR only	−	−	+		
3.	(Random Specimen) Alpha Naphthol Phosphate + Fast Blue RR	+	−	+	9.2	77
	Fast Blue RR	−	−	−		
4.	(Random Specimen) Alpha Naphthol Phosphate + Fast Blue RR	+	−	+	6.2	62
	Fast Blue RR only	−	−	−		
5.	(Random Specimen) Alpha Naphthol Phosphate + Fast Blue RR	+	−	+	3.7	70
	Fast Blue RR	−	−	−		

*Ultrafast liver

Figure 1: A 61 year old female with cirrhosis

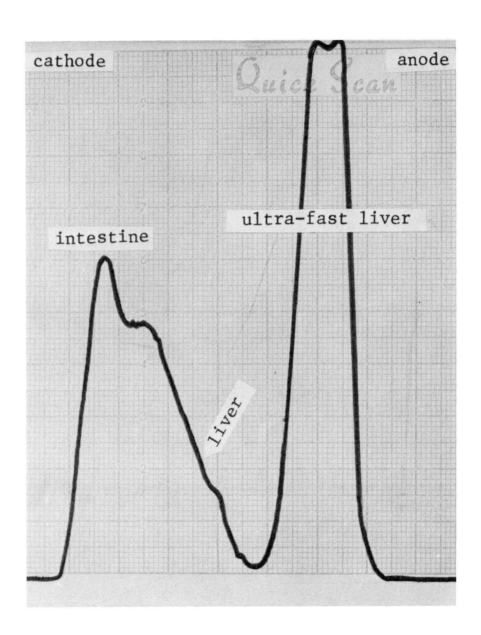

Figure 2: A 72 year old female with carcinoma of the head of the pancreas with metastasis to the liver.

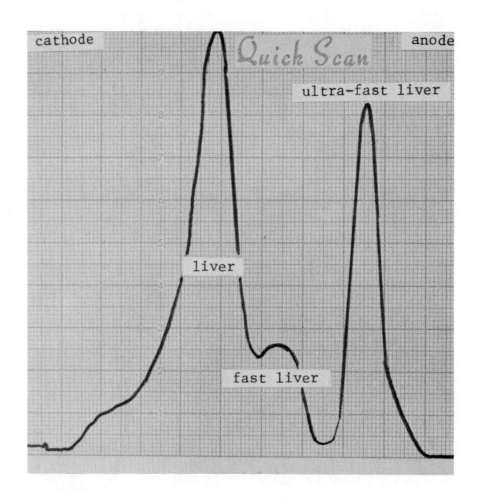

Figure 3: A 60 year old female with metastatic ovarian carcinoma.

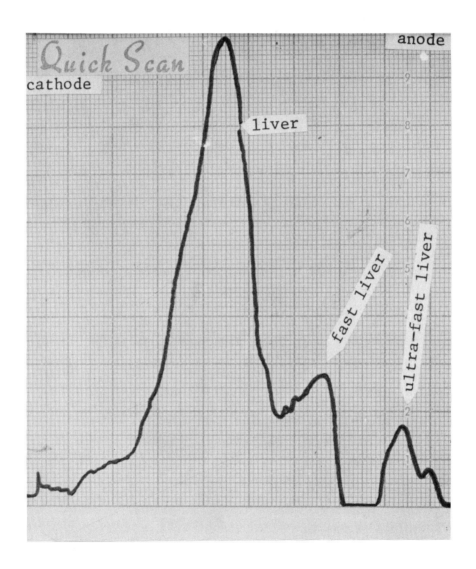

Figure 4: A 60 year old female with metastatic breast carcinoma.

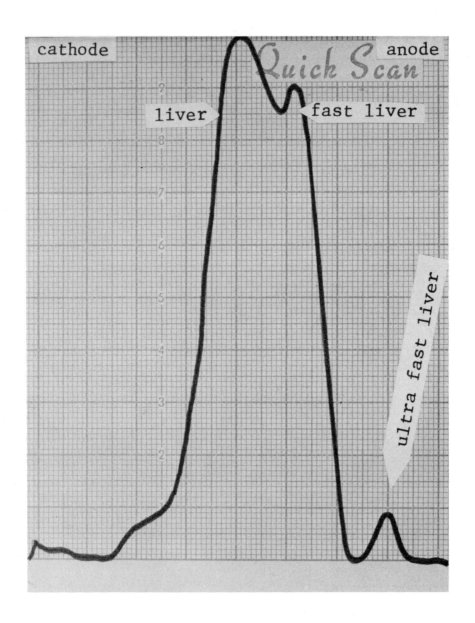

Figure 5: A 55 year old female with liver-spleen scan consistent with cirrhosis; multiple areas of isotope uptake on bone scan.

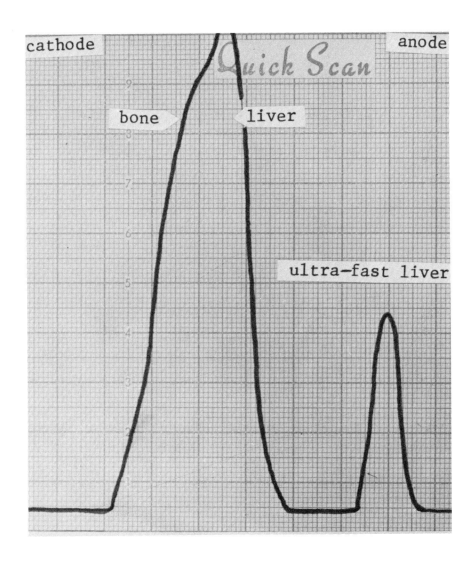

Figure 6: A 55 year old female with abnormal liver function tests believed to be secondary to allopurinol and/or colchicine--resolution on withdrawal of drugs.

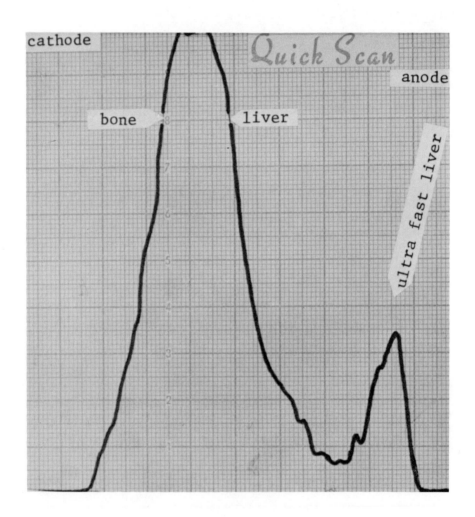

Figure 7: A 31 year old male with Stage IV Hodgkin's disease' filling defect on liver-spleen scan.

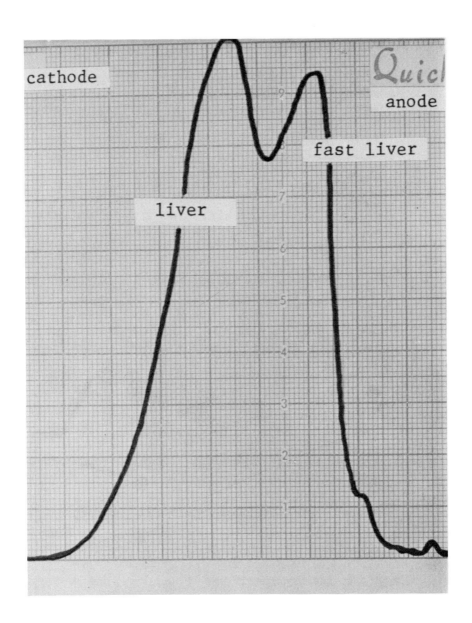

Figure 8: A 67 year old female with hemochromatosis; discrete filling defect on liver-spleen scan.

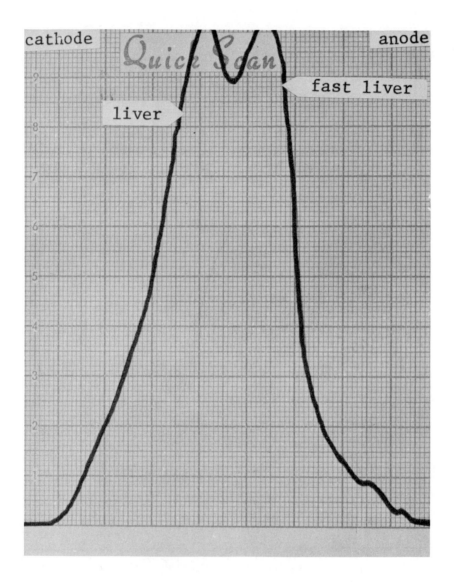

Figure 9: A 52 year old male with renal adenocarcinoma

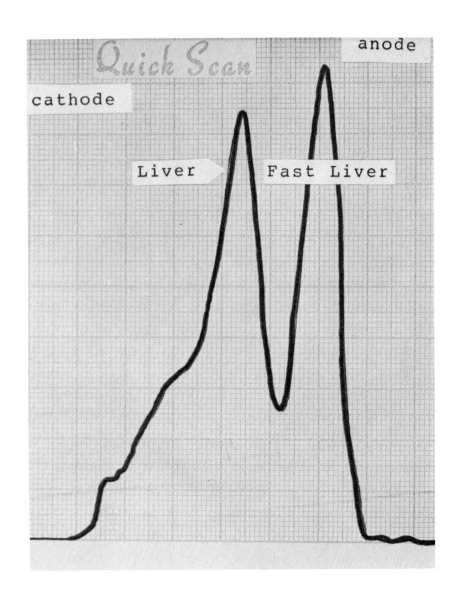

Figure 10: A 60 year old male with metastatic adenocarcinoma.

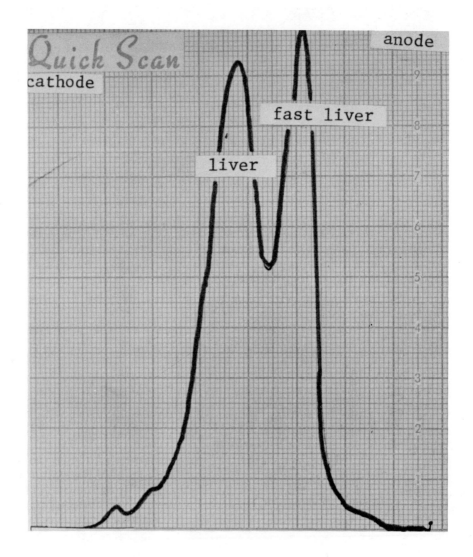

THE BREAST CANCER PROGNOSTIC STUDY: ANALYZING THE METASTATIC POTENTIAL OF HUMAN BREAST CANCERS

Marvin A. Rich and Michael J. Brennan

Scientific, Pathology and Surgical Associates of the Michigan Cancer Foundation, Breast Cancer Prognostic Study, Michigan Cancer Foundation, Detroit, Michigan 48201

Long term survival of the breast cancer patient is determined by the probability of metastatic spread and subsequent recurrence of the disease. It is likely that the events associated with metastatic proliferation depend on specific and measurable characteristics of a patient and her tumor.

With the active participation of our clinical community, we organized the Breast Cancer Prognostic Study aimed at the detailed characterization of large numbers of primary human breast tumors and their hosts, and the identification of those factors associated with early recurrence and metastatic disease.

The experimental approach to this problem is based on the observation that human breast cancers can be graded using a simple cytological criteria (Bloom and Richardson) and that the grade is a moderately accurate indicator of the metastatic potential of that tumor. It had been suggested earlier, and confirmed by our own observations, that morphological grade, an accurate predictor of early recurrence, is an intrinsic characteristic of each tumor and that recurrences in the breast or at distant sites recapitulate the cytological features (and grade) of the primary tumor.

Procedures for acquisition of specimens, medical history, and clinical findings have been developed. Currently, three to five new patients per week are being entered into this study. Maintenance of contact with the patient and her physicians for clinical follow-up and specimen collection are part of the regularized system. In our laboratory research units, we have established assays to characterize both the tumors and hosts with respect to their morphological, endocrinological, immunological, and biochemical nature. The tests have been chosen for their predicted or proven association with malignancy and metastatic potential. The progress in these areas and the identification of biological and biochemical characteristics which can be used to predict the recurrence of breast cancers will be discussed.

THE LIBRARY
UNIVERSITY OF CALIFORNIA
San Francisco